Lamiaceae Species

Lamiaceae Species

Biology, Ecology and Practical Uses

Special Issue Editor

Milan Stankovic

MDPI • Basel • Beijing • Wuhan • Barcelona • Belgrade • Manchester • Tokyo • Cluj • Tianjin

Special Issue Editor
Milan Stankovic
University of Kragujevac
Serbia

Editorial Office
MDPI
St. Alban-Anlage 66
4052 Basel, Switzerland

This is a reprint of articles from the Special Issue published online in the open access journal *Plants* (ISSN 2223-7747) (available at: https://www.mdpi.com/journal/plants/special_issues/lamiaceae_species).

For citation purposes, cite each article independently as indicated on the article page online and as indicated below:

LastName, A.A.; LastName, B.B.; LastName, C.C. Article Title. *Journal Name* **Year**, *Article Number*, Page Range.

ISBN 978-3-03928-418-4 (Pbk)
ISBN 978-3-03928-419-1 (PDF)

Cover image courtesy of Milan Stankovic.

© 2020 by the authors. Articles in this book are Open Access and distributed under the Creative Commons Attribution (CC BY) license, which allows users to download, copy and build upon published articles, as long as the author and publisher are properly credited, which ensures maximum dissemination and a wider impact of our publications.

The book as a whole is distributed by MDPI under the terms and conditions of the Creative Commons license CC BY-NC-ND.

Contents

About the Special Issue Editor vii

Preface to "Lamiaceae Species: Biology, Ecology and Practical Uses" ix

Noura S. Dosoky and William N. Setzer
The Genus *Conradina* (Lamiaceae): A Review
Reprinted from: *Plants* **2018**, *7*, 19, doi:10.3390/plants7010019 1

Vjera Bilušić Vundać
Taxonomical and Phytochemical Characterisation of 10 *Stachys* Taxa Recorded in the Balkan Peninsula Flora: A Review
Reprinted from: *Plants* **2019**, *8*, 32, doi:10.3390/plants8020032 11

Carmen X. Luzuriaga-Quichimbo, José Blanco-Salas, Carlos E. Cerón-Martínez, Milan S. Stanković and Trinidad Ruiz-Téllez
On the Possible Chemical Justification of the Ethnobotanical Use of *Hyptis obtusiflora* in Amazonian Ecuador
Reprinted from: *Plants* **2018**, *7*, 104, doi:10.3390/plants7040104 25

Yasmina Benabdesslem, Kadda Hachem, Khaled Kahloula and Miloud Slimani
Ethnobotanical Survey, Preliminary Physico-Chemical and Phytochemical Screening of *Salvia argentea* (L.) Used by Herbalists of the Saïda Province in Algeria
Reprinted from: *Plants* **2017**, *6*, 59, doi:10.3390/plants6040059 33

Katerina Tzima, Nigel P. Brunton and Dilip K. Rai
Qualitative and Quantitative Analysis of Polyphenols in *Lamiaceae* Plants—A Review
Reprinted from: *Plants* **2018**, *7*, 25, doi:10.3390/plants7020025 45

Gabriela Almada-Taylor, Laura Díaz-Rubio, Ricardo Salazar-Aranda, Noemí Waksman de Torres, Carla Uranga-Solis, José Delgadillo-Rodríguez, Marco A. Ramos, José M. Padrón, Rufina Hernández-Martínez and Iván Córdova-Guerrero
Biological Activities of Extracts from Aerial Parts of *Salvia pachyphylla* Epling Ex Munz
Reprinted from: *Plants* **2018**, *7*, 105, doi:10.3390/plants7040105 75

Leila Bendifallah, Rachida Belguendouz, Latifa Hamoudi and Karim Arab
Biological Activity of the *Salvia officinalis* L. (Lamiaceae) Essential Oil on *Varroa destructor* Infested Honeybees
Reprinted from: *Plants* **2018**, *7*, 44, doi:10.3390/plants7020044 89

Theofilos Mailis and Helen Skaltsa
Polar Constituents of *Salvia willeana* (Holmboe) Hedge, Growing Wild in Cyprus
Reprinted from: *Plants* **2018**, *7*, 18, doi:10.3390/plants7010018 101

Duangjai Tungmunnithum, Laurine Garros, Samantha Drouet, Sullivan Renouard, Eric Lainé and Christophe Hano
Green Ultrasound Assisted Extraction of *trans* Rosmarinic Acid from *Plectranthus scutellarioides* (L.) R.Br. Leaves
Reprinted from: *Plants* **2019**, *8*, 50, doi:10.3390/plants8030050 113

About the Special Issue Editor

Milan Stankovic is an Associate Professor of Plant Science at the Department of Biology and Ecology, Faculty of Sciences, University of Kragujevac, Republic of Serbia. He completed his Ph.D. in Botany at the same University and postdoctoral research at the Université François-Rabelais de Tours, France. He is the Head of the Department of Biology and Ecology. Dr. Stankovic has published over 200 references including articles in peer-reviewed journals, edited books, book chapters, conference papers, meeting abstracts etc. He is an editor, editorial board member, and reviewer for several scientific journals. He currently works as an Associate Editor of the *Plants* journal.

Preface to "Lamiaceae Species: Biology, Ecology and Practical Uses"

Lamiaceae (Labiatae) is an important plant family that consists of 250 genera and more than 7000 species. The largest genera that belong to this family are *Salvia*, *Scutellaria*, *Stachys*, *Plectranthus*, *Hyptis*, *Teucrium*, *Thymus*, *Vitex*, *Nepeta*, etc. A large number of Lamiaceae species inhabit different ecosystems and have a great diversity with a cosmopolitan distribution. Most of the species are aromatic and possess a complex mixture of bioactive compounds that contribute to overall biological activity in both in vitro and in vivo conditions. Secondary metabolites with potent antioxidant, anti-inflammatory, antimicrobial, antiviral, and anticancer effects are crucial in terms of the previously mentioned biological activities. Moreover, plants that belong to this family are valuable in food, cosmetics, flavoring, fragrance, perfumery, pesticide, and pharmaceutical industries. Because of a wide range of applications, the plants of the Lamiaceae family are widely cultivated and are, therefore, regarded as an indispensable source of functional food. Based on these facts, numerous research works have been carried out on different aspects of Lamiaceae species in regard to its biology, ecology and applications. The Special Issue Book entitled "Lamiaceae Species: Biology, Ecology and Practical Uses" contributes to the knowledge of selected Lamiaceae species from several aspects, such as diversity and phytogeography, taxonomy, ethnobotany, quantitative, and qualitative composition as well as biological activity of secondary metabolites. I am grateful to all the authors for their contributions as well as the reviewers for their professional suggestions and decisions. I am highly thankful to the *Plants* MDPI team for many years of collaboration. Especially, I would like to give special thanks to Sylvia Guo and Shuang Zhao.

Milan Stankovic
Special Issue Editor

Review

The Genus *Conradina* (Lamiaceae): A Review

Noura S. Dosoky [1] and William N. Setzer [1,2,*]

1. Aromatic Plant Research Center, 615 St. George Square Court, Suite 300, Winston-Salem, NC 27103, USA; ndosoky@aromaticplant.org
2. Department of Chemistry, University of Alabama in Huntsville, Huntsville, AL 35899, USA
* Correspondence: wsetzer@chemistry.uah.edu; Tel.: +1-256-824-6519

Received: 24 February 2018; Accepted: 10 March 2018; Published: 11 March 2018

Abstract: *Conradina* (Lamiaceae) is a small genus of native United States (US) species limited to Florida, Alabama, Mississippi, Tennessee and Kentucky. Three species of *Conradina* are federally listed as endangered and one is threatened while two are candidates for listing as endangered. The purpose of the present review is to highlight the recent advances in current knowledge on *Conradina* species and to compile reports of chemical constituents that characterize and differentiate between *Conradina* species.

Keywords: *Conradina*; essential oil; ursolic acid; cytotoxicity; antimicrobial; antileishmanial; phylogenetic analysis

1. Introduction

Conradina A. Gray (Lamiaceae) is a small genus of morphologically distinctive, narrow-leaved, minty-aromatic shrubs endemic to the southeastern United States of America (USA) [1]. It consists of six to nine US native species. According to the Plant List database [2] and Integrated Taxonomic Information Species (ITIS), the acceptable number of species for the genus is only six (*Conradina canescens* A. Gray, *C. cygniflora* C.E. Edwards, Judd, Ionta & Herring, *C. etonia* Kral & McCartney, *C. glabra* Shinners, *C. grandiflora* Small and *C. verticillata* Jennison) while *C. brevifolia* Shinners, *C. montana* Small and *C. puberula* Small are considered synonyms of other species. The Missouri Botanical Garden, however, lists nine distinct species [3]. *Conradina* species are characterized by very dense hairs on their lower leaf surfaces and by a sharply bent corolla tube in the flowers [1,4]. Asa Gray established the genus *Conradina* in 1870, named for the American botanist Solomon White Conrad [5]. *Conradina* species grow well in xeric habitats with well-drained sandy soils. It is thought that *Conradina* may be a pioneer species in disturbed areas since it has the ability to colonize xeric disturbed habitats [1,5].

Each species of *Conradina* occupies a separate geographical region [1,6]. Five of the species are endemic to Florida, one is native to west Florida, south Alabama, and south Mississippi, and one is endemic to north-central Tennessee and Kentucky (Figure 1). Due to habitat destruction from residential, commercial, and agricultural land conversions and their very restricted distribution, three of these species are on the United States Fish and Wildlife Service (USFWS) federal list as endangered (*Conradina brevifolia*, *C. glabra*, and *C. etonia*) [5], one as threatened (*C. verticillata*) [6] and two species are considered candidates for listing as endangered (*C. grandiflora* and *C. cygniflora*).

The aim of the present review is to summarize the literature in order to document secondary metabolites that characterize and differentiate between *Conradina* species. We shall also assess the reported biological activities of *Conradina* species, with particular focus on their phylogenetic affinities and anti-microbial, anti-leishmanial and cytotoxic properties.

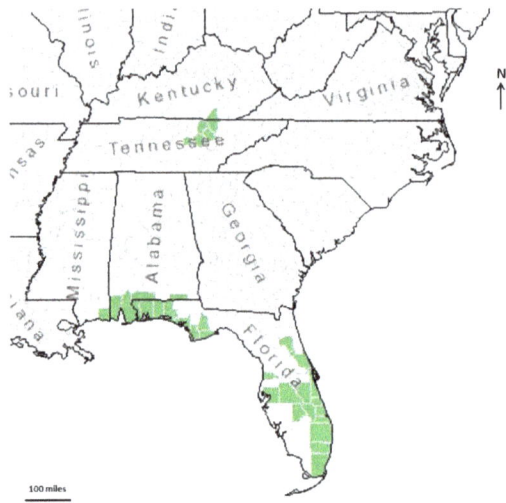

Figure 1. Locations of known populations of *Conradina* species.

2. Description

Conradina represents a group of evergreen, compact, perennial shrubs with virgate branches. Leaves are aromatic and needle-like with narrow, entire, revolute leaf-blades. Flowers are solitary or in a group of few in axillary cymes. The calyx is two-lipped; the lower lip has two long, narrow lobes while the upper lip has three short broad lobes. Corolla is white to purple in color, two-lipped; the lower lip has three lobes and usually dotted while the upper lip is erect and slightly concave.

Conradina canescens A. Gray (false rosemary) is the only species of this genus that is relatively common in its range and the most morphologically variable (Table 1). It is native to a small area of west Florida and south Alabama and southern parts of Mississippi [7]. It occurs on sunny, dry sand soils on coastal dunes, in sand scrubs and in dry longleaf pineland ecosystems. It shows tolerance for the heat and humidity of the Southeast and is considered a drought-tolerant landscape plant that grows best in lean, sandy soils [8]. *C. canescens* is one of the indicator species for wooded dunes [9] and is an excellent candidate for restoration of coastal areas. It can be used for beach projects that require planting on the back side of primary dunes, or any side of secondary dunes. It has also been used in gardens as a groundcover as the branches can reach up to 1 m in spread. False rosemary was easily propagated in the University of Florida Institute of Food and Agricultural Sciences Extension using softwood stem cuttings from terminal shoots taken during the growing season.

Morphologically, *C. brevifolia* Shinners (short-leaved rosemary) is very similar to *C. canescens*, differing by its shorter leaves and greater number of flowers per axil, and it is also similar to the endangered *C. glabra* [1,5]. *C. brevifolia* was described as a new species by Shinners [4]. Many taxonomic reviews of *Conradina* have upheld *C. brevifolia* as a distinct species yet it is taxonomically questionable [1,6,10,11]. Some taxonomists are still uncertain that *C. brevifolia* is a true different species and therefore they treat it as a synonym to *C. canescens*. *C. brevifolia* has a very restricted range in the middle of the Lake Wales Ridge in Central Florida [12–14] and has been listed as a federally endangered species in 1993 [5]. Habitat destruction for agricultural and residential purposes is the main threat to this species.

C. etonia Kral & McCartney (Etonia rosemary) was discovered in the Etonia Scrub in Putnam County, Florida in September 1990 and described as a new species in 1991 [6]. *C. etonia* is similar to *C. grandiflora* in general appearance and the flowers are large and quite similar but the leaves of *C. etonia* are broader than *C. grandiflora* and have clearly visible lateral veins on the lower surface [1,15].

C. etonia is found in very limited areas of deep white sand scrub dominated by sand pines (*Pinus clausa*) and scrubby oaks (*Quercus* spp.) on dry soils. Its entire known range is within a subdivision of Putnam County containing streets and a few residences, which make it the most narrowly distributed species of *Conradina* [6]. *C. etonia* was listed as a federally endangered species in July 1993 [5]. The latest survey of Etonia rosemary counted about eight subpopulations on Etonia Creek State Forest containing fewer than 1000 total individuals [16]. The major threats facing *C. etonia* include habitat loss due to development projects for residential housing, horticultural collection, hurricane damage and invasive species or lack of natural processes, such as fire [17,18]. Because *C. etonia* requires light but can tolerate the light shade of openings the in scrub, fire suppression is one of the major limiting factors as it results in over-shading by overstory scrub vegetation rendering the habitat less favorable for *C. etonia* [18].

C. grandiflora Small (large-flowered rosemary) is endemic to the scrub habitat of Florida's east coast between Daytona Beach and Miami as well as near Orlando and Okeechobee County [5]. A regime of frequent fires benefits *C. grandiflora* because it cannot tolerate shade and requires an open sunny habitat. *C. grandiflora* is a candidate for listing as an endangered species but of a lower priority than other *Conradina* species [5]. Loss of the Florida scrub habitat is the major threat to this species.

C. glabra Shinners (Apalachicola rosemary) was described in a sand hill community east of the Apalachicola River in Liberty County, Florida in 1962 [1,4]. This species is very narrowly distributed and restricted to Liberty County, Florida. Out of its seven known locations, six *C. glabra* subpopulations (artificially divided by the Florida gas transmission pipeline) are remaining, all of which are located on private silvicultural lands and subject to much disturbance. Its other historical locations were converted to pine plantations, and can no longer support this species. Because of homozygosity and inbreeding, many of flowers of *C. glabra* are male sterile, and the stamens are very malformed [1]. It was added to the federal list of endangered species in 1993 [5,19,20], primarily due to habitat loss and modification of forestry practices. There is no record of how abundant *C. glabra* was before its habitat was altered because the silviculture operations began before its discovery.

C. verticillata Jennison (Cumberland rosemary) is the only species of this genus that is not found in Florida. The species was established in 1933 [21]. It is a rare species, existing only as three small subpopulations in Tennessee and one subpopulation in Kentucky [22]. *C. verticillata* is found only in boulder bars, cobble bars, sand bars and gravel bars in close proximity to rivers on the Cumberland Plateau of north-central Tennessee and southeastern Kentucky [22]. The preferred habitat conditions include open to slightly-shaded area, moderately deep well-drained sandy soil, periodic flooding, and topographic features like narrow channels or depressions on gravel bars. *C. verticillata* is the only species that is triploid ($n = 3$) and accordingly has a lower ability to reproduce and disperse sexually and greatly reduced seed germination rates. The fresh seeds of *C. verticillata* are physiologically dormant and require cold stratification to germinate [23]. It has been on the federal list of threatened species since 1991 [24]. The major threats facing *C. verticillata* are habitat destruction, general deterioration of water quality, competition and shading by woody plants.

In 2009, analysis of patterns of genetic structure based on microsatellites in *Conradina* led to the identification of *C. cygniflora* Edwards, Judd, Ionta & Herring [25]. *C. cygniflora* can only be found in Dunns Creek state park in south-central Putnam County, Florida. It occupies nine tightly-clustered sites that probably form around two to four self-sustaining subpopulations. *C. cygniflora* carries several unique morphological characters that distinguish this species from all other described *Conradina* species including thin-walled unicellular hairs, epidermis features, and larger calyces. Due to its exceptionally limited geographic distribution and low number of individuals, *C. cygniflora* is considered a candidate for listing as federally endangered [25].

There is very little information in the literature about *C. montana* and *C. puberula*. *C. montana* Small was reported in sandy woods and ravines, Appalachian plateau near Rugby, Tennessee while *C. puberula* Small was observed in pine-lands, northern Gulf coast region, Florida [26,27]. Some taxonomic discrepancies surround *C. montana* and *C. puberula* since some experts treat both species as synonyms of *C. verticillata* and *C. canescens*, respectively.

Table 1. Morphological characteristics of *Conradina* species.

Species	Morphological Characteristics	References
C. canescens	Small shrub, up to 1 m high. Leaves are 7 to 20 mm long, mostly longer than the internodes. Leaf blades are pubescent on both sides. One to three flowers per axil, lower corolla-lip 8–10-mm long; lateral lobes longer than wide. Calyx-tube hirsute or villiou-hirsute.	[5]
C. brevifolia[a]	Small shrub, up to 1 m high. Leaves are short fleshy 6.0 to 8.2 mm long, mostly shorter than the internodes, covered with short downy hairs and many tiny glands on the upper side. One to six lavender flowers per axil.	[5,11]
C. etonia	Straight slender shrub, about 1.5 m high. Leaves have hairy, veiny, glandular blades 1.5–3 cm long and 3–9 mm wide with tightly rolled edges. Three to seven flowers per axil. Pink to lavender in color with darker dotted lower petal.	[5,15,18]
C. glabra	Small shrub, about 80 cm high but some individuals reach up to 2 m. Leaves are opposite, up to 1.5 cm long, hairless on the upper surface. Two to three flowers per axil. Corolla is 1.5–2 cm long, white to pale lavender in color with a band of purple dots on the lower lip.	[5,19,20]
C. grandiflora	Erect shrub, 1.5–2.0 m high, with hairy branches and twigs. Leaves are hairy, glandular, up to 1.5 cm long. Year-round hairy lavender flowers with darker lavender spots, lower lip is 12–15 mm long with lateral lobes longer than wide. This species has the largest flowers of genus *Conradina*.	[5]
C. verticillata	Erect shrub, 0.5 m high with reclining branches. Leaves are about 2.5 cm long, very narrow, and arranged in tight bunches that appear as whorls around the stems. Flowers are 2.5 cm long, purple to white and borne in leaf-like clusters of bracts at the ends of the stems.	[24]
C. montana[b]	Short shrub less than 0.5 m high with diffuse branches. Leaves are narrowly linear, 5–16 mm long. Leaf blades are glabrous on the upper surface. Minute flowers with corolla 3.5–4 mm long. Calyx-tube hirsutulous.	[23]
C. puberula[a]	Short shrub of about 0.5 m high with numerous slender branches. Leaves are narrowly linear but strongly revolute and clavate, 12–20 mm long. Calyx-tube minutely canescent, 5–7 mm long. Flowers appear in racemes of 2–6 per axil, with corolla 4–5 mm long.	[26,27]
C. cygniflora	Virgate shrub up to 1 m high, branches are erect to spreading, internodes 5–43 mm long. Leaves persistent, appearing fascicled- verticillate; narrowly obovate, 9–33 mm long. The abaxial leaf surface is densely-covered by simple unicellular hairs. Cymes carry 1–5 subsessile flowers, densely pubescent, 1.3–12.5 mm long. Large calyx of 8.5–11 mm long; densely covered with simple hairs, upper lip upcurved, 3.6–4.4 mm long, lower lip 4.3–5.5 mm long. Corolla strongly bilabiate, 20–29 mm long, lavender, shading to white in throat, with purple spots; abaxial surface of upper lip darker lavender.	[25]

[a] *C. brevifolia* and *C. puberula* are sometimes treated as synonyms to *C. canescens*, [b] *C. montana* is sometimes thought as a synonym to *C. verticillata*.

3. Phylogenetic Studies

Conradina is a member of Lamiaceae and belongs to a clade of New World Mentheae. The closely related genera that share similar morphology and habitat preference with *Conradina* include *Dicerandra* spp., *Piloblephis* spp., *Stachydeoma* spp., and the woody, southeastern U.S. species of *Clinopodium* such as *C. ashei, C. georgianum, C. coccineum,* and *C. dentatum* [28]. Despite the fact that *Conradina* species are morphologically distinguishable, there are two problems about species status based on morphology. First, *C. brevifolia* (endangered species) shares a lot of morphological characteristics with *C. canescens* (relatively widespread in its range). Second, populations of *Conradina* in Santa Rosa County, Florida, have morphological characteristics in common with both *C. glabra* (endangered species) and *C. canescens*. Due to the endangered status of four *Conradina* species, it is important to resolve the species relationships, define species limits and elucidate the evolutionary processes in order to help with conservation plans. *Conradina* was subjected to several phylogenetic analyses using four data partitions: plastid DNA, internal transcribed spacer (ITS), and two members of the GapC gene family [29–31]. These analyses did not resolve species relationships in *Conradina*. In agreement morphological evidence, ITS results supported a monophyletic *Conradina* while plastid DNA results did not. Some *Conradina* populations showed moderate levels of inbreeding, but inbreeding does not seem as a major factor. Because previous analyses failed to resolve species relationships separately, combined analyses were carried out but the results still did not resolve the relationships within *Conradina*. Studies using genotype data from quickly evolving microsatellite loci suggested that both ancient interspecific hybridization and incomplete lineage sorting of ancestral polymorphism have likely occurred in *Conradina* and that based on the patterns of genetic structure (which corresponds to species boundaries), *Conradina* species have diverged genetically from one another (nonmonophyly) despite having unique distribution patterns and distinguishing morphological characteristics [31].

4. Essential Oils and Non-Volatile Components

4.1. Essential Oils

The essential oils obtained from the aerial parts of *Conradina* spp. are rich in terpenes, terpenic aldehydes and ketones, and terpenic alcohols [32,33]. Terpenes released from *Conradina* are allelopathic, and are believed to help prevent wildfires [34,35] and to protect against insects [32]. The main components quantified in the essential oils of *Conradina* species with the respective percentages (%) are summarized in Table 2. The essential oil of *C. cygniflora* has not been studied yet. 1,8-Cineole and camphor are the major components of *C. canescens*, *C. brevifolia*, *C. glabra* and *C. verticillata* [7]. The common components found in all *Conradina* species (excluding *C. cygniflora*) share a terpenoid skeleton but vary in cyclization and oxygenation (Figure 2). These common components include 1,8-cineole (bicyclic monoterpene ether), camphor (bicyclic monoterpene ketone), α-pinene (bicyclic monoterpene), β-pinene (bicyclic monoterpene), β-myrcene (acyclic monoterpene), borneol (bicyclic monoterpene alcohol), sabinene (bicyclic monoterpene), camphene (bicyclic monoterpene), and β-caryophyllene (bicyclic sesquiterpene).

Table 2. Chemical composition of essential oils of *Conradina* species.

Species	Major Oil Components (%)	Unique Component(s)	Reference
C. brevifolia [a]	Camphor (9.7–17.54%) and 1,8-cineole (1.97–4.86%)	α- and β-farnesene	[7]
C. canescens [a]	Camphor (0.27–23.64%), 1,8-cineole (0.17–3.34%), cis-pinocamphone (0–8.74%)	none	[7]
C. canescens [b]	1,8-cineole (5.2–25.2%), camphor (5.7–8.0%), α-pinene (3.2–5.6%), *p*-cymene (3.3–5.9%), cis-pinocamphone (1.3–5.5%), myrtenal (5.2–8.1%), myrtenol (3.4–9.2%), verbenone (4–4.5%), and myrtenyl acetate (5.0–5.4%)	-	[33]
C. etonia [a]	Camphor (30.55–35.65%), limonene (3.77–6.33%), camphene (2.92–3.75%), and β-caryophyllene (2.95–6.54%)	β-elemene, 4-carene and α-terpineol	[32]
C. glabra [a]	1,8-cineole (2.38–7.34%) and camphor (11.78–15.88%)	Dolcymene and bornyl acetate	[7]
C. grandiflora [a]	β-pinene (4.38–5.81%) and β-cubebene (1.95–6.56%)	Calarene and β-pinone	[7]
C. verticillata [a]	1,8-cineole (3.15–3.78%) and camphor (5.81–8.35%)	Germacrene B and 2,5,6-trimethyl-1,3,6-heptatriene	[7]
C. cygniflora	Data not available	N/A	N/A

[a] Essential oil obtained by solvent extraction, [b] essential oil obtained by hydrodistillation.

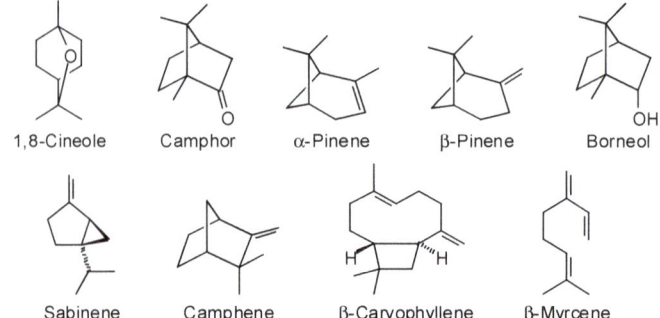

Figure 2. Chemical structures of common components found in six species of *Conradina*.

4.2. Nonvolatile Components

Conradina is a reservoir of several important bioactive molecules. However, the chemical studies on *Conradina* are very limited in number. Six compounds were separated from the chloroform extract of *C. canescens* leaves (Figure 3): ursolic acid, betulin, β-amyrin, myrtenic acid, *n*-tetracosane, and oleanolic acid [36]. *C. canescens* leaves are considered a natural source for ursolic acid [37].

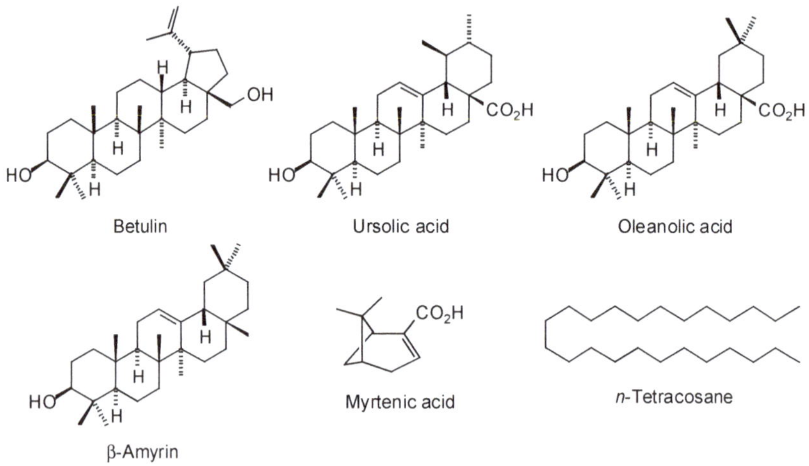

Figure 3. Chemical structures of nonvolatile components isolated from *Conradina canescens*.

5. Biological Activities

5.1. Allelopathic Activity

The term allelopathy describes the chemical interactions between plants where one plant interferes with the germination and growth of another plant. Allelopathic compounds are thought to enter the surrounding environment via volatilization, leaching with rain, and decomposition of plant litter, thereby inhibiting the growth of competitors or species that may threaten the plant's survival. In this manner, the plant may have an ecological role in its ecosystem by affecting plant spacing, succession, and community composition [35]. Terpenoids, especially monoterpenoids, are considered one of the 14 allelochemical classes [38,39] that play an important role in the interactions between plants [40,41]. *C. canescens* is thought to play a significant ecological role in maintaining a healthy scrub ecosystem by inhibiting the germination of native grasses [42]. It was suggested that by inhibiting germination of the grasses, the development of the more fire-prone sand hill ecosystem is prevented, and the scrub ecosystem is maintained [35]. Because detailed knowledge of allelopathic actions in natural plant communities can provide excellent models for new strategies in developing highly selective herbicides, the allelopathic effect of *C. canescens* has been a subject of several studies [34,35,43]. Water washes of fresh *C. canescens* leaves were reported to have strong inhibitory effects on germination and growth of *Schizachyrium scoparium* [34,43]. The essential oil of *C. canescens* and purified ursolic acid remarkably inhibited the germination of *Lactuca sativa* and *Lolium perenne* [37]. The phytotoxic activity of *C. canescens* oil can be attributed to the high concentration of 1,8-cineole which can inhibit the seed germination and seedling growth of lettuce in a dose-dependent manner [44–48] by strongly inhibiting mitochondrial respiration and all stages of mitosis [49]. The presence of significant quantities of ursolic acid suggests that this compound might contribute to the strong allelopathic effect of *C. canescens* [50,51]. Ursolic acid is thought to act as a natural detergent by leading water-insoluble monoterpenes to form micelles, rendering them water-soluble, thereby enhancing their ability to leach into rainwater for delivery into the soil [42,50]. Ursolic acid helps to co-solubilize the allelopathic monoterpenes in water to be more effective [52].

5.2. Antibacterial and Antifungal Activity

The essential oil of *C. canescens* has no antibacterial activity but has a slight antifungal activity against *Botrytis cinerea* [33] while the crude extracts showed good antifungal activity against *Botrytis cinerea* [36]. Ursolic acid, the major component in *C. canescens* leaves was reported to have selective antibacterial activity against *S. aureus* [36,53] while its isomer, oleanolic acid, was active against *S. aureus* and *Pseudomonas aeruginosa*. Interestingly, isolated *n*-tetracosane showed antibacterial activity against *Staphylococcus aureus* and *S. epidermidis* and antifungal activity against *Aspergillus niger* and *B. cinerea*. Isolated myrtenic acid was active against *S. aureus*, *A. niger*, *B. cinerea* and *Candida albicans* [36].

5.3. Cytotoxic Activity

The essential oil of *C. canescens* has no significant in vitro cytotoxic activity against the human breast tumor cell lines MCF-7 and MDA-MB-231 [33]. The crude chloroform extract of *C. canescens* showed a significant cytotoxic activity against MCF-7 (human breast tumor), MDA-MB-231 (human breast tumor) and 5637 (human bladder tumor) cell lines which is attributed to the presence of betulin and ursolic acid, which when tested individually showed strong effects against all tested cell lines [36].

5.4. Antileishmanial Activity

The crude extracts have promising antileishmanial activity against promastigotes and intracellular amastigotes of *Leishmania amazonensis*. However, the dichloromethane extract had some cytotoxicity on the host cells. The antileishmanial activity of the nonpolar extract was attributed to the presence of ursolic acid and betulin being the major constituents [36].

6. Conclusions

This review summarizes the current status, distribution, chemical value and biological studies on genus *Conradina*. There is still some debate about the exact number of *Conradina* species as well as the classification/taxonomy of this genus and more studies, especially genetic studies, are still needed. The available information increases the importance of protecting *Conradina* from extinction. Future studies on this genus may provide valuable information on the mechanism of evolution.

Acknowledgments: This work was carried out as part of the activities of the Aromatic Plant Research Center (APRC, https://aromaticplant.org/). The authors are grateful to dōTERRA International (https://www.doterra.com/US/en) for financial support of the APRC.

Author Contributions: N.S.D conceived, collected the references, and wrote the initial draft of the review; W.N.S. edited the review.

Conflicts of Interest: The authors declare no conflicts of interest.

References

1. Gray, T.C. *A Monograph of the Genus Conradina*; Vanderbilt University: Nashville, TN, USA, 1965.
2. The Plant List. Available online: http://www.theplantlist.org/tpl1.1/search?q=conradina (accessed on 8 March 2018).
3. The Missouri Botanical Garden. Available online: http://www.tropicos.org/NameSearch.aspx?name=Conradina&commonname= (accessed on 8 March 2018).
4. Shinners, L.H. Synopsis of *Conradina* (Labiatae). *Sida* **1962**, *1*, 84–88.
5. U.S. Fish and Wildlife Service, (USFWS). Endangered and Threatened Wildlife and plants: Endangered or threatened status for five Florida plants. *USFWS Fed. Regist. Rules Regul.* **1993**, *58*, 37432–37443.
6. Kral, R.; McCartney, R.B. A new species of *Conradina* (Lamiaceae) from northeastern peninsular Florida. *Sida* **1991**, *14*, 391–398.
7. Peterson, C.L. *Analysis of the Essential Oils, Leaf Ultrastructure, and the In Vitro Growth Response of the Mint Genus Conradina*; Florida Institute of Technology: Melbourne, FL, USA, 1998.
8. Harrison, M. *Groundcovers for the South*; Pineapple Press: Sarasota, FL, USA, 2006.
9. Looney, P.B.; Gibson, D.J. The relationship between the soil seed bank and above-ground vegetation of a coastal barrier island. *J. Veg. Sci.* **1995**, *6*, 825–836. [CrossRef]
10. Wunderlin, R.P.; Richardson, D.; Hansen, B. *Status Report on Conradina brevifolia*; U.S. Fish and Wildlife Service: Jacksonville, FL, USA, 1980.
11. Kral, R. *A Report on Some Rare, Threatened, or Endangered Forest-Related Vascular Plants of the South*; USDA Forest Service: Atlanta, GA, USA, 1983.
12. Christman, S.P.; Judd, W.S. Notes on plants endemic to Florida scrub. *Fla. Sci.* **1990**, *53*, 52–73.
13. Christman, S.P. *Endemism and Florida's Interior Sand Pine Scrub*; Final Report No. GFC-84-101; Florida Game and Fresh Water Fish Commission, Nongame Wildlife Program: Tallahassee, FL, USA, 1988.
14. U.S. Fish and Wildlife Service (USFWS). *Short-Leaved Rosemary (Conradina brevifolia): Five-Year Review*; USFWS: Atlanta, GA, USA, 2008.
15. U.S. Fish and Wildlife Service (USFWS). *Recovery Plan for Etonia rosemary (Conradina etonia)*; USFWS: Atlanta, GA, USA, 1994.
16. U.S. Fish and Wildlife Service, (USFWS). *Conradina etonia, Five-Year Review*; USFWS: Atlanta, GA, USA, 2007.
17. Peterson, C.L.; Weigel, R.C. In vitro propagation of *Conradina etonia*. *Fla. Sci.* **2002**, *65*, 201–207.
18. U.S. Fish and Wildlife Service, (USFWS). *Etonia rosemary (Conradina etonia)*; USFWS: Atlanta, GA, USA, 2005.
19. U.S. Fish and Wildlife Service, (USFWS). *Recovery Plan for Apalachicola rosemary (Conradina glabra)*; USFWS: Atlanta, GA, USA, 1994.
20. U.S. Fish and Wildlife Service, (USFWS). *Conradina glabra (Apalachicola rosemary): Five-Year Review*; USFWS: Atlanta, GA, USA, 2009.
21. Jennison, H.M. A new species of *Conradina* from Tennessee. *J. Elisha Mitchell Sci. Soc.* **1933**, *48*, 268–269.
22. U.S. Fish and Wildlife Service, (USFWS). *Recovery Plan for Nineteen Central Florida Scrub and High Pineland Plants (Revised)*; USFWS: Atlanta, GA, USA, 1996.
23. Albrecht, M.A.; Penagos, J.C.Z. Seed germination ecology of three imperiled plants of rock outcrops in the southeastern United States. *J. Torrey Bot. Soc.* **2012**, *139*, 86–95. [CrossRef]
24. U.S. Fish and Wildlife Service, (USFWS). Endangered and Threatened Wildlife and plants: *Conradina verticillata* (Cumberland rosemary) determined to be threatened. *USFWS Fed. Regist. Rules Regul.* **1991**, *56*, 60937–60941.
25. Edwards, C.E.; Judd, S.; Ionta, G.M.; Herring, B. Using population genetic data as a tool to identify new species: *Conradina cygniflora* (Lamiaceae), a new, endangered species from Florida. *Syst. Bot.* **2009**, *34*, 747–759. [CrossRef]
26. Small, J.K. *Manual of the Southeastern Flora*; The University of North Carolina Press: Chapel Hill, NC, USA, 1933.
27. Small, J.K. Studies in the botany of the Southeastern United States-XIV. *Bull. Torrey Bot. Club* **1898**, *25*, 469–484. [CrossRef]

28. Trusty, J.L.; Olmstead, R.G.; Bogler, D.J.; Guerra, A.S.; Ortega, J.F. Using molecular data to test a biogeographic connection of the Macaronesian genus Bystropogon (Lamiaceae) to the New World: A case of conflicting phylogenies. *Syst. Bot.* **2014**, *29*, 702–715. [CrossRef]
29. Edwards, C.E.; Soltis, D.E.; Soltis, P.S. Molecular phylogeny of *Conradina* and other scrub mints (Lamiaceae) from the southeastern USA: Evidence for hybridization in pleistocene refugia? *Syst. Bot.* **2006**, *31*, 193–207. [CrossRef]
30. Edwards, C.E.; Lefkowitz, D.; Soltis, D.; Soltis, P. Phylogeny of *Conradina* and related Southeastern scrub mints (Lamiaceae) based on GapC gene sequences. *Int. J. Plant Sci.* **2008**, *169*, 579–594. [CrossRef]
31. Edwards, C.E.; Soltis, D.E.; Soltis, P.S. Using patterns of genetic structure based on microsatellite loci to test hypotheses of current hybridization, ancient hybridization and incomplete lineage sorting in *Conradina* (Lamiaceae). *Mol. Ecol.* **2008**, *17*, 5157–5174. [CrossRef] [PubMed]
32. Quinn, B.P.; Bernier, U.R.; Booth, M.M. Identification of compounds from Etonia rosemary (*Conradina etonia*). *J. Chromatogr. A* **2007**, *1160*, 306–310. [CrossRef] [PubMed]
33. Dosoky, N.S.; Stewart, C.D.; Setzer, W.N. Identification of essential oil components from *Conradina canescens*. *Am. J. Essent. Oils Nat. Prod.* **2014**, *2*, 24–28.
34. Williamson, G.B.; Fischer, N.H.; Richardson, D.R.; de la Peña, A. Chemical inhibition of fire-prone grasses by fire-sensitive shrub, *Conradina canescens*. *J. Chem. Ecol.* **1989**, *15*, 1567–1577. [CrossRef] [PubMed]
35. Fischer, N.H.; Williamson, G.B.; Weidenhamer, J.D.; Richardson, D.R. In search of allelopathy in the Florida scrub: The role of terpenoids. *J. Chem. Ecol.* **1994**, *20*, 1355–1380. [CrossRef] [PubMed]
36. Dosoky, N.S.; Moriarity, D.M.; Setzer, W.N. Phytochemical and biological investigations of *Conradina canescens* Gray. *Nat. Prod. Commun.* **2016**, *11*, 25–28. [PubMed]
37. Dosoky, N.S. *Isolation and Identification of Bioactive Compounds from Conradina canescens Gray*; University of Alabama in Huntsville: Huntsville, AL, USA, 2015.
38. Rice, E.L. *Allelopathy*, 2nd ed.; Academic Press: Orlando, FL, USA, 1984.
39. Macias, F.A.; Molinillo, J.M.G.; Varela, R.M.; Galindo, J.C.G. Allelopathy-a natural alternative for weed control. *Pest Manag. Sci.* **2007**, *63*, 327–348. [CrossRef] [PubMed]
40. Khan, M.A.; Ungar, I.A. The effect of salinity and temperature on the germination of polymorphic seeds and growth of *Atriplex triangularis* Willd. *Am. J. Bot.* **1984**, *71*, 481–489. [CrossRef]
41. Muller, W.H.; Muller, C.H. Volatile growth inhibitors produced by *Salvia* species. *Bull. Torrey Bot. Club* **1964**, *91*, 327–330. [CrossRef]
42. De la Peña, A.C. *Terpenoids from Conradina canescens (Labiatae) with Possible Allelopathic Activity*; Louisiana State University: Baton Rouge, LA, USA, 1985.
43. Fischer, N.H.; Tanrisever, N.; Williamson, G.B. Allelopathy in the Florida scrub community as a model for natural herbicide actions. In *Biologically Active Natural Products: Potential Use in Agriculture*; Cutler, H.G., Ed.; American Chemical Society: Washington, DC, USA, 1988.
44. Qiu, X.; Yu, S.; Wang, Y.; Fang, B.; Cai, C.; Liu, S. Identification and allelopathic effects of 1,8-cineole from *Eucalyptus urophylla* on lettuce. *Allelopath. J.* **2010**, *26*, 255–264.
45. Nishida, N.; Tamotsu, S.; Nagata, N.; Saito, C.; Sakai, A. Allelopathic effects of volatile monoterpenoids produced by *Salvia leucophylla*: Inhibition of cell proliferation and DNA synthesis in the root apical meristem of *Brassica campestris* seedlings. *J. Chem. Ecol.* **2005**, *31*, 1187–1203. [CrossRef] [PubMed]
46. Zunino, M.P.; Zygadlo, J.A. Effect of monoterpenes on lipid oxidation in maize. *Planta* **2004**, *219*, 303–309. [PubMed]
47. Singh, H.P.; Batish, D.R.; Kohli, R.K. Allelopathic effect of two volatile monoterpenes against bill goat weed (*Ageratum conyzoides* L.). *Crop Prot.* **2002**, *21*, 347–350. [CrossRef]
48. Abrahim, D.; Braguini, W.L.; Kelmer-Bracht, A.M.; Ishii-Iwamoto, E.L. Effects of four monoterpenes on germination, primary root growth, and mitochondrial respiration of maize. *J. Chem. Ecol.* **2000**, *26*, 611–624. [CrossRef]
49. Macias, F.A.; Galindo, J.C.G.; Molinillo, J.M.G. *Allelopathy: Chemistry and Mode of Action of Allelochemicals*, 1st ed.; CRC Press: Boca Raton, FL, USA, 2003.
50. Rizvi, S.J. *Allelopathy: Basic and Applied Aspects*; Springer Science & Business Media: Dordrecht, The Netherlands, 2012.
51. Szakiel, A.; Grzelak, A.; Dudek, P.; Janiszowska, W. Biosynthesis of oleanolic acid and its glycosides in *Calendula officinalis* suspension culture. *Plant Physiol. Biochem.* **2003**, *41*, 271–275. [CrossRef]

52. Weidenhamer, J.D.; Macias, F.A.; Fischer, N.H.; Williamson, G.B. Just how insoluble are monoterpenes? *J. Chem. Ecol.* **1993**, *19*, 1799–1807. [CrossRef] [PubMed]
53. Wolska, K.I.; Grudniak, A.M.; Fiecek, B.; Kraczkiewicz-Dowjat, A.; Kurek, A. Antibacterial activity of oleanolic and ursolic acids and their derivatives. *Cent. Eur. J. Biol.* **2010**, *5*, 543–553. [CrossRef]

© 2018 by the authors. Licensee MDPI, Basel, Switzerland. This article is an open access article distributed under the terms and conditions of the Creative Commons Attribution (CC BY) license (http://creativecommons.org/licenses/by/4.0/).

Review

Taxonomical and Phytochemical Characterisation of 10 *Stachys* Taxa Recorded in the Balkan Peninsula Flora: A Review

Vjera Bilušić Vundać

University of Zadar, Department of Health Studies, Splitska 1, 23000 Zadar, Croatia; vjerab_2000@yahoo.com

Received: 21 September 2018; Accepted: 22 January 2019; Published: 29 January 2019

Abstract: The genus *Stachys* is one of the largest genera of the Lamiaceae, and it comprises about 300 species. Some species are highly polymorphic, with a number of infraspecific taxa. The aim of the present review is to summarise the available knowledge on 10 taxa belonging to the Balkan Peninsula flora (*S. alpina* L., *S. germanica* L., *S. menthifolia* Vis., *S. obliqua* Waldst. Et Kit., *S. officinalis* (L.) Trevis., *S. palustris* L., *S. recta* L. subsp. *recta*, *S. recta* L. subsp. *subcrenata* (Vis.) Briq., *S. salviifolia* Ten., and *S. sylvatica* L.) in order to enable insight into the identified biologically active substances and their possible application in intrageneric differentiation.

Keywords: Lamiaceae; *Stachys*; essential oil; fatty acids; amino acids; flavonoids; iridoids; phytochemical; chemotaxonomy

1. Introduction

Stachys L. (woundwort) comprises approximately 300 species, and is considered one of the largest genera of Lamiaceae, with nearly worldwide distribution. Most of these species occur in the warm temperate regions of the Mediterranean and southwest Asia, with secondary centres in North and South America and southern Africa; woundwort is not found in Australia and New Zealand [1]. The genus comprises 19 taxa in the flora of Croatia and neighbouring areas (Domac, 1973), among which eight species are endemic to the Balkan Peninsula or even narrower regions [2]. This article comprises phytochemical and taxonomical characterisation of 10 taxa: *S. alpina* L., *S. germanica* L., *S. menthifolia* Vis., *S. obliqua* Waldst. Et Kit., *S. officinalis* (L.) Trevis., *S. palustris* L., *S. recta* L. subsp. *recta*, *S. recta* L. subsp. *subcrenata* (Vis.) Briq., *S. salviifolia* Ten., and *S. sylvatica* L. among the presented species. Presented data also include chemotaxonomical guidance for highly polymorphic species of *S. recta*, with a proposal for the differentation of two of the presented subspecies [1–7]. *S. menthifolia* Vis. is not mentioned by Domac, but is listed in Flora Europaea as endemic species for Balkan Peninsula, and has been identified on several localities of Croatia and Montenegro [3,8,9]. The aim of present review is to find a scientific literature that contains research on the biologically active compounds of presented *Stachys* taxa belonging to the Balkan Peninsula flora and to report the findings on the chemotaxonomical composition that could help to differentiate within the *Stachys* spp. Presented data also include chemotaxonomical guidance for highly polymorphic species of *S. recta*, with a proposal for the differentation of two presented subspecies [1–7,10].

2. Taxonomy and Systematics

Genus *Stachys* is composed of annual or perennial herbs or subshrubs that are hispid or soft-pubescent. Leaves are sesille or stipulate, blade oblong to ovate, with serrate to crenate margins. Flowers are sessile or short-stalked, with two or more clustered in the axils of the leaves on the upper part of the stem. The calyx is bell-shaped, and has five lobes or teeth. Corolla has a narrow tube, generally with a short, pouched spur on the lower side of the tube. The upper lip is erect or generally parallel to tube axis, concave, entire (notched), and generally hairy, and the lower lip is perpendicular to the tube axis or reflexed, three (two)-lobed, and glabrous to hairy. The nutlets are oblong to ovoid [2–7,10].

S. alpina L. (alpine woundwort) is a green, hirsute, glandular perrenial up to 100 cm in height. It grows in shadow on limestone soils from mountainous to subalpine areas [2,3,10] [Figure 1].

Figure 1. *Stachys alpina* L. (**a**) plant in flowering stage; (**b**) detail of the flower; (**c**) detail of leaf.

S. officinalis L. (betony) is a perennial herb of 15–60 cm height, with sparse hairs, a short woody rhizome, well-marked basal rosettes, and erect, simple, or slightly branched stems. Betony occurs in open woods, stony grassy places, and dry meadows from the sea level up to 1600 m [2,3,10,11] [Figure 2].

S. palustris L.(marsh woundwort) is a sparsely to densely hairy perennial plant of shady spots in marshes, bogs, ponds, lakes, and stream margins, damp places, roadside verges, and as a weed in cultivated fields growing up to 120 cm from a creeping rhizome [2,3,10] [Figure 3].

Figure 2. *S. officinalis* L. (**a**) plant in flowering stage; (**b**) detail of the flower; (**c**) detail of leaf.

Figure 3. *S. palustris* L. (**a**) plant in flowering stage; (**b**) detail of the flower; (**c**) detail of leaf.

S. recta L. subsp. *recta* (perennial yellow woundwort) is erect or ascending, subglabrous to sparsely hirsute, and usually an eglandular plant of up to 100 cm in height, which grows on different dry and stony habitats [2,3,10] [Figure 4].

Figure 4. *S. recta* L. *subsp. recta* (**a**) plant in flowering stage; (**b**) detail of the flower; (**c**) detail of leaf.

S. recta L. subsp. *subcrenata* (Vis.) Briq. has a very similar appearance and distribution as compared to the *S. recta* L. *subsp. Recta*, but has narrower leaves and often a glandular calyx with unequal teeth [2,3,10,11] [Figure 5].

S. sylvatica L. (hedge woundwort) is an erect and hirsute perennial found in shady spots in woodland, forests, roadsides, alpine meadows, and grasslands, growing up to 120 cm. It differs from marsh woundwort in habitat preference, the broader leaves, and its characteristic, unpleasant smell when crushed, but the two can hybridise where they co-exist [2,3,10,11] [Figure 6].

Figure 5. *S. recta* L. *subsp. subcrenata* (**a**) plant in flowering stage; (**b**) detail of the flower; (**c**) detail of leaf.

Figure 6. *S. sylvatica* L. (**a**) plant in flowering stage; (**b**) detail of the flower; (**c**) detail of leaf.

S. obliqua Waldst. et Kit. is an erect lanate-hirsute perennial with oblong-lanceolate leaves that is native to the Balkan Peninsula [12] [Figure 7].

S. germanica L. (downy woundwort, german hedgenettle) is a densely tomentose and ascendent or erect biennial or perennial with well-marked basal rosettes, growing up to 120 cm and found in light forests, shrubby sunny slopes, rocks, and cut-over areas [11,12] [Figure 8].

S. salviifolia Ten. (Mediterranean woundwort) is a densely tomentose or lanate-tomentose perennial growing up to about 80 cm that is widely distributed on rocky places in the littoral areas [2,3,10,11] [Figure 9].

Figure 7. *Stachys obliqua* Waldst. et Kit. (by courtesy of Saxifraga Foundation).

Figure 8. *Stachys germanica* L. (by courtesy of Saxifraga Foundation).

Figure 9. *S. salviifolia* Ten. (**a**) plant in flowering stage; (**b**) detail of the flower; (**c**) detail of leaf.

Stachys menthifolia Vis. is an endemic species that was first described by Roberto Visiani (1829) at the locality of Kotor (Montenegro), who also identified this species in habitat near Dubrovnik (Croatia) [13]. According to the Flora Europaea, the species is endemic of the Balkan Peninsula, and spread in Albania, Greece, and Yugoslavia [3]. In 2002, Šilić and Šolić identified new localities in the eastern part of the Biokovo Massif [9]. *S. menthifolia* Vis. is a calciphilic plant, with the distinctive feature of a short, glandulary, and pilose stem, and is mostly present in the fissures of limestone or in the open limy rocky grounds, as in the Croatian localities. In the majority of the Montenegrin localities, the plants grow individually or in small groups, often in the strata of thermophilic woods and the underbrush of pubescent oak (*Quercetalia pubescentis*), in stone fissures, or in open surfaces, while the habitats of the new Croatian localities are mainly open limy rocky grounds with denser populations composed of larger individuals [9,14,15].

3. Phytochemical Characteristics

3.1. Essential Oils

Similar to many other representatives of the family *Lamiaceae*, *Stachys* species produce essential oil. Essential oils are complex mixtures and in spite of the large size of the genus *Stachys*, the composition of the volatile compounds is known in only a small number of species.

Several studies refer to the composition of the essential oil within the members of the present review: *S. alpina* [1,10], *S. officinalis* [1,16–18], *S. sylvatica* [1,16,17,19–22], *S. palustris* [1,23], *S. germanica* [16,17,24], *S. menthifolia* [14,15,25], *S.obliqua* [26–28], *S. recta* [1,17,25,29–31], and *S. salviifolia* [1,10].

Isocaryophyllene and b-caryophyllene, which are dominating constituents of the previously investigated essential oil from one population of *S. officinalis* in Montenegro [16], were totally absent in that from a population in Serbia, with germacrene D being the main compound [17]. Another study conducted in Serbia also confirmed sesquiterpene hydrocarbons as being the main fraction and identified germacrene D, β-caryophyllene, and α-humulene as the main constituents [18].

In a study conducted on samples from Croatia, germacrene D and (E)-caryophyllene were the main compounds of *S. officinalis*, with isocaryophyllene being totally absent [1].

Concerning the prior investigations of *S. sylvatica*, it is noteworthy that germacrene D is present in a large amount in the sample from Italy [19], and was also present in high proportions in the essential oil from Serbia [17], while the monoterpene fraction is present in all of the samples in a very low percentage only. *S. sylvatica* essential oil from Croatia was also rich in germacrene D, but in contrast to the oil from Italy and Serbia, monoterpene hydrocarbons were found in a large amount with α-pinene and β-pinene being the most abundant constituents of this fraction. The investigation of the chemical composition of essential oils from *S. sylvatica* collected from three different wild populations in Kosovo [20] showed that the leaves and flowers of *S. sylvatica* were characterised by three main constituents: α-pinene, β-pinene, and germacrene D, which is aligned with the findings on the main constituents that were identified in a study conducted in Croatia. Essential oil obtained from *S. sylvatica* sample collected in Turkey contained high proportions of germacrene D, α-pinene and β-caryophyllene, but the β-pinene constituent was not identified [21]. Another study conducted on a *S. sylvatica* sample collected in Turkey also implicated monoterpene hydrocarbons as the major constituents of essential oil, but with difference in a composition, with γ-muurolene, α-cedrene (11.2%), and limonene being the most abundant components [22].

Sesquiterpene hydrocarbons were the main group of essential oil constituents of *S. salviifolia* and *S. palustris* sample from Croatia, with germacrene D being the main constituent of *S. salviifolia*, in contrast to *S. palustris*, which contained a significant aldehyde fraction and a high amount of alcohols [1]. The essential oil from the aerial parts of *S. palustris* from both samples from Italy was characterised mainly by carbonylic compounds, fatty acids, and their esters, along with sesquiterpenoidic compounds and phenols [23,24].

The composition of the essential oil from one population of *S. recta* from Serbia [17] was similar to the oil from a population in Turkey [31] by the preponderance of 1-octen-3-ol. In contrast to the oil from one population in Greece [25] and to the oil from Turkey, in the oil from Serbia, linalool was present in very low percentage only. An analysis of *S. recta* oil from another population in Serbia revealed a different composition compared to other oils; germacrene D and E-caryophyllene were the main constituents of the essential oil, while 1-octen-3-ol was present in very low proportions, and linalool was totally absent [17]. The reason for these variations in the chemical composition of the essential oil was probably because there are different subspecies of *S. recta*, which can differ in their essential oil composition.

Two subspecies of *S. recta* (*S. recta* subsp. *recta* and *S. recta* subsp. *subcrenata*) that were analysed in the Croatian study were collected at the contiguous habitats. Although both subspecies had a high amount of sesquiterpene hydrocarbons, their composition differed; germacrene D was the main component of *S. recta* subsp. *subcrenata* essential oil, but its level is low in the *S. recta* subsp. *recta* oil. *S. recta* subsp. *subcrenata* was also characterised by a high percentage of oxygenated sesquiterpenes, whereas *S. recta* subsp. *recta* had a very low amount of these compounds. On the other hand, *S. recta* subsp. *recta* was rich in aldehydes, and had a rather high amount of ketones and fatty acids, which were present in the essential oil of *S. recta* subsp. *subcrenata* in a very low percentage only [1].

Stachynone and stachynene are constituents of the essential oil of *S. officinalis* and *S. recta*, as reported by Maly (1985), and were absent in all of the essential oils that were studied. One reason may be that these compounds decompose in the gas chromatographic injector [29].

Sesquiterpene hydrocarbons were the main group of constituents found in *S. menthifolia*, as demonstrated for several other *Stachys* species [25]. Abietatriene and 13-epi-manoyl oxide were present in high amounts in *S. menthifolia* from Greece [25], while samples collected in two different habitats in Croatia (Biokovo and Dubrovnik) differed in their essential oil composition. The major constituent found in the essential oil from the sample collected in Dubrovnik was, as in the sample from Greece, diterpenoid abietatriene (11.7%), with significant amounts of sesquiterpene hydrocarbons α-bisabolene (8.4%) and β-caryophyllene (7.4%). Oxygenated sesquiterpenes were the most abundant volatiles in the sample from Biokovo, followed by diterpene hydrocarbons. The content of sesquiterpene hydrocarbons was much higher in the sample from Dubrovnik, which also showed no presence of 4'-methoxyacetophenone. The authors attributed the differences in chemical oil composition to the different harvesting dates of the samples [14,15].

The *S. alpina* sample from Croatia showed the presence of high amounts of oxygenated sesquiterpenes, with a significant aldehyde fraction portion [1]. *S. germanica* essential oil samples from Serbia and Italy also showed oxygenated sesquiterpenes as the main constituents, with a sample from Serbia having a large amount of monoterpene hydrocarbons as well (11,3%) [17,24].

A similar composition was found for *S. obliqua* samples collected in Turkey, which had a significant portion of monoterpene hydrocarbons and a very high percentage of oxygenated sesquiterpenes with germacrene D as the main constituent [26–28].

A detailed overview of the main components (sesquiterpenes and monoterpenes) in the essential oil of *Stachys* taxa reviewed is presented in Tables 1 and 2.

Table 1. Main component groups found in *Stachys* taxa essential oil from aerial parts (monoterpene hydrocarbons: MH; oxygenated monoterpenes: OM; sesquiterpene hydrocarbons: SH; oxygenated sesquiterpenes: OS).

Taxa	Origin	MH	OM	SH	OS	References
S. alpina	Croatia	/	2.5	10.7	28.0	[1]
	Serbia	12.3	0.8	37.8	4.8	[17]
S. germanica	Italy, locality Parco Nazionale del Pollino	0.5	0.6	20.6	4.5	[24]
	Italy, locality Parco Nazionale delle Madonie	1.3	5.1	10.3	2.9	[24]

Table 1. Cont.

Taxa	Origin	MH	OM	SH	OS	References
S. menthifolia	Croatia, locality Ploče	0.3	1.7	6.4	48.4	[14]
	Croatia, locality Dubrovnik	1.5	4.4	35.1	24.0	[14]
	Croatia, locality Biokovo	0.1–4.6	1.7–2.9	1.4–6.4	48.4–58.9	[14]
	Croatia, locality Vrgorac, alt. 180 meter supra mare	0.1	2.9	1.4	58.9	[14]
	Croatia, locality Vrgorac, alt. 30-130 meter supra mare	4.6	2.4	4.2	48.4	[14]
	Greece	4.3	0.5	15.3	14.2	[25]
S. obliqua	Turkey	10.3	/	72.2	4.1	[26]
S. officinalis	Croatia	2.4	0.3	70.4	18.7	[1]
	Serbia	0.3	0.4	71.1	7.7	[17]
	Serbia	0.8	3.7	69.1	14.8	[18]
S. palustris	Croatia	/	/	22.7	16.2	[1]
	Italy					[23]
	Italy	0.5	3.1	5.4	10.6	[24]
S. recta	Croatia (subs. recta)	4.1	2.9	22.2	2.1	[1]
	Croatia (subsp. subcrenata)	/	6.4	46.5	24.4	[1]
	Serbia	15.8	/	68.7	4.1	[17]
	Greece	1.6	41.9	6.6	/	[25]
	Serbia	2.2	/	9.3	4.6	[30]
	Turkey	17.7	29.7	13.9	/	[31]
S. sylvatica	Croatia	38.5	0.8	41.7	7.7	[1]
	Serbia	3.05	/	70.83	11.45	[17]
	Italy	1.7	/	72.6	6.6	[19]
	Turkey	48.9	1.2	32.1	2.6	[22]
S. salviifolia	Croatia	5.3	0.8	58.4	14.9	[1]

Table 2. Main sesquiterpenes and monoterpenes recorded in *Stachys* taxa (percentige above 5% presented).

Compound	Taxa	Origin	%	References
(E)-Caryophyllene	S. officinalis	Croatia	14.6	[1]
		Serbia	14.1	[18]
	S. sylvatica	Croatia	9.9	[1]
		Turkey	20.8	[21]
	S. sylvatica (leaves)	Italy	31.7	[19]
	S. recta subsp. recta	Croatia	5.4	[1]
	S. menthifolia	Croatia, locality Dubrovnik	7.4	[14]
	S. officinalis	Montenegro	22.9	[16]
Caryophyllene oxide	S. palustris	Croatia	16.2	[1]
		Italy	7.8	[23]
		Italy	7.8	[24]
	S. menthifolia	Croatia, locality Ploče	5.2	[14]
	S. officinalis	Montenegro	6.5	[16]
Germacrene D	S. obliqua	Croatia	20.1	[1]
	S. officinalis	Turkey	45.3	[26]
		Serbia	19.9	[18]
	S. recta subsp. subcrenata	Croatia	19.7	[1]
	S. salviifolia	Croatia	22.3	[1]
	S. sylvatica	Croatia	13.6	[1]
		Turkey	23.9	[21]
	S. sylvatica (inflorescence)	Italy	55.2	[19]
	S. sylvatica (leaves)	Italy	31.7	[19]
	S. obliqua	Turkey	8.2	[27]
		Turkey	6.2	[28]

Table 2. Cont.

Compound	Taxa	Origin	%	References
α-Humulene	S. officinalis	Serbia	7.5	[18]
α-Pinene	S. sylvatica	Croatia	21.4	[1]
		Turkey	19.6	[21]
β-Pinene	S. sylvatica	Croatia	12.3	[1]
Linalool	S.recta	Turkey	13.0	[31]
(−)-β-Linalool	S.recta	Greece	33.9	[25]
(E)-Nerolidol	S. alpina	Croatia	12.3	[1]
γ-Cadinene	S. recta subsp. recta	Croatia	6.9	[1]
δ-Cadinene	S. sylvatica (leaves)	Italy	31.7	[19]
α-Cadinol	S. recta subsp. subcrenata	Croatia	9.5	[1]
β-Elemene	S. salviifolia	Croatia	9.4	[1]
α-Bisabolene	S. menthifolia	Croatia, locality Dubrovnik	7.4	[14]
Valeranone	S. menthifolia	Croatia, locality Vrgorac, alt. 30-130 meter supra mare	5.7	[14]
8-α-Acetoxyelemol	S. menthifolia	Croatia, locality Vrgorac, alt. 180 meter supra mare	21.3	[14]
		Croatia, locality Vrgorac, alt. 30-130 meter supra mare	6.9	
		Croatia, locality Ploče	12.8	
Limonene	S. obliqua	Turkey	6.2	[28]
	S. sylvatica	Turkey	37.0	[22]
Thymol	S. obliqua	Turkey	6.2	[28]
	S. palustris	Italy	5.8	[23]
α-Cedrene	S. sylvatica	Turkey	37.0	[22]
γ-Muurolene	S. sylvatica	Turkey	10.2	[22]

3.2. Other Biologically Active Substances

Similar to most of the plants belonging to the Lamiaceae family, the *Stachys* taxa have been submitted to several investigations in order to determine the content of the biologically active compounds. These investigations have reported the presence of flavonoids, phenolic acids, iridoids, and terpenoids.

Bilušić Vundać et al. investigated seven Croatian Stachys taxa (*S. alpina*, *S. officinalis*, *S. palustris*, *S. recta* subsp. *recta*, *S. recta* subsp. *subcrenata*, *S. salviifolia*, and *S. sylvatica*) in order to determine the content of the polyphenols, tannins, phenolic acids, and flavonoids, as well as the composition of flavonoids and phenolic acids. All of the taxa that were tested had a high content of total polyphenols, a medium content of total phenolic acids, and a rather low content of tannins and flavonoids. The most present phenolic compound in the investigated samples was found to be the chlorogenic acid (it was not present only in *S. recta* subsp. *recta*), and in samples of *S. recta* subsp. *recta* and *S. sylvatica*, the presence of isoquercitrin, luteolin-7-O-glucoside, and quercitrin were determined. Rutin was determined in samples of *S. recta* subsp. *recta*, *S. recta* subsp. *subcrenata*, *S. sylvatica*, and *S. salviifolia*. Although *S. recta* subsp. *recta* didn't contain chlorogenic acid, it was found to be the richest in flavonoids whose presence was determined [32,33].

Karioti et al. (2010) determined 14 flavonoid glycosides in *S. recta*, with the majority of the constituents being the derivatives of isoscutellarein, and there being only four apigenin-*p*-coumaroyl derivatives [34].

Tundis et al. (2014) reviewed all of the available data on the biologically active substances in *Stachys* taxa, and found the following diterpenes to be present in the taxa that are included in this review: annuanone, stachylone, and stachone (*S. palustris, S. sylvatica*); abietal, abietatriene, stachysic acid, scareol, *trans*-Phytol, 6β-hydroxy-ent-kaur-16-ene, 6b,18-dihydroxy-*ent*-kaur-16-ene, kaur-16-ene,13-epi-manoyl oxide, and betolide (*S. sylvatica*); betonicoside A-D and *trans*-Phytol (*S. officinalis*); abietatriene, abieta-8, 11, 13-trien-7-one, 13-epi-manoyl oxide, *cis*-Phytol, and Labd-13-en-8,15-ol and kaur-16-ene (*S.menthifolia*); abietatriene and epi-Laurenene (*S. germanica*); and *cis*-Phytol, 11α,18-dihydroxy-*ent*-kaur-16-ene, kaur-16-ene, diacetyl derivatives, and monoacetyl derivatives of stachysolone (*S. recta*). Furthermore, in their overview of data on iridoids isolated in *Stachys* taxa, the following compounds were determined for the taxa presented: *S. recta* and *S. alpina* contained harpagide, ajugoside, aucubin, acetylharpagide, and harpagoside; *S. germanica* contained harpagide, aucubin, acetylharpagide, and harpagoside, *S. palustris* contained harpagide and aucubin; while *S. alpina* contained harpagide, ajugoside, aucubin, and harpagoside [35]. Furthermore, recent research studies of the active substances in *S. palustis* identified not only iridoids harpagide and 8-O-acetyl-harpagide, but also monomelittoside, which was the first identification of this compound in *S. palustris* species [36].

Bilušić Vundać (2006) identified several amino acids in *Stachys* species: aspartic acid (*S. alpina, S. sylvatica, S. officinals,* and *S. salviifolia*), asparagine (*S. sylvatica, S. alpina,* and *S. recta* subsp. *subcrenata*), γ-aminobutyric acid were determined in all of the investigated taxa (*S. alpina, S. officinalis, S. palustris, S. recta* subsp. *recta, S. recta* subsp. *subcrenata, S. salviifolia,* and *S. sylvatica*), as well as alanine, leucine, and serine [37].

4. Discussion and Conclusions

The presented studies revealed some differences between *Stachys* taxa, indicating the existence of a chemical polymorphism. The taxonomic relationships within the taxa are complex, and have been the objective of several investigations. Although different types of biologically active substances were determined within the investigated *Stachys* species, the main differentiation is based on essential oil and fatty acid composition. Tundis et al. (2014) focused on taxonomy within *Stachys* and *Betonica* subgenera, as well as infrageneric differentiation within *Stachys* species focusing on these chemotaxonomical markers. Based on the collected data on essential oil, three groups of *Stachys* species were observed: species mainly characterised by monoterpene hydrocarbons (to which *S. sylvatica* was assigned), species mainly characterised by oxygenated monterpenes (to which *S. recta* was assigned), and species that contain similar amounts of sesquiterpene hydrocarbons and oxygenated sesquiterpenes [35].

The essential oil approach, including its fatty acid composition, was taken by Bilušić Vundać (2006), who through principal components (PCA) and hierarchical cluster analyses (CA) differentiated seven *Stachys* taxa to two distinct groups, first comprising *S. alpina, S. recta* subsp. *Recta*, and *S. palustris*, and second comprising *S. officinalis, S. sylvatica, S. recta* subsp. *subcrenata*, and *S. salviifolia*. The species within the first group showed a high content of aldehyde components, and in the case of *S. alpina* and *S. palustris*, a high content of alcohol, while in the species within the second group, germacrene D was identified as the main component.

The essential oil and fatty acid approach was also used in chemotaxonomic differentiation by Kiliç at al. (2017), who defined germacrene D/β-caryophyllene the essential oil chemotype and 6-octadecanoic acid fatty acid chemotype for *S. sylvatica* [21].

The chemical investigations of *Stachys* species have determined the presence of monterpenes, sesquiterpenes, diterpenes, iridoids, phenylethanoids, flavonoids, and fatty acids as the main biological constituents. The content of these components and their occurrence in investigated species is variable, due to many external factors that impact the secondary metabolite composition of the plant species [1,35].

The differentiation of the presented *Stachys* species based on the available data on their chemical composition requests further chemotaxonomical and molecular research in order to present a key for infrageneric classification. Througout history, basic plant species and subspecies identification has relied mostly on morphological traits. For morphological characterisation, taxonomists have formulated a descriptor list, such as leaf shape and colour, flower colour, etc. With the development of scientific methods, chemotaxonomy research increased in order to enable separation within the taxa based on the presence and quantity of a specific chemical compound [38]. In recent studies, the authors have tried to resolve the classification of *Stachys* taxa based on the prevalent essential oil compositions from the aerial parts [10,21,26]. Nevertheless, the results vary, since different geographic localities, seasons, harvest periods, properties of soils, and climatic conditions strongly affect the secondary metabolite composition of plant species, especially their essential oil composition [1,21,35]. Therefore, traditional evaluation methods and chemotaxonomy should be combined with molecular markers that are not influenced by the environmental factors, for a better distinction among *Stachys* taxa.

Taking all of the above into consideration, a potential suggestion for differentiation can be suggested within the presented investigations only for two *S. recta* subspecies, which were collected at identical conditions and habitat (time of collection, harvest period, soil, and climate), which could support the opinion of Visiani (1829), who described *S. recta* subsp. *subcrenata* as separate species and named it *S. subcrenata* [10].

Although many constituents of this genus have been already identified, further investigations of *Stachys* spp. are required in order to clearly identify the chemotaxonomical markers for specific taxa and clarify their taxonomic relationships.

Funding: This research received no external funding.

Acknowledgments: The author would like to thank Saxifraga Foundation (http://www.saxifraga.nl) for providing photos of *S. obliqua* (photo by Jasenka Topić) and *S. germanica* (photo by Jan Willem Jongepier).

Conflicts of Interest: The author declares no conflict of interest.

References

1. Bilušić Vundać, V.; Pfeifhofer, H.W.; Brantner, A.H.; Maleš, Ž.; Plazibat, M. Essential oil of seven *Stachys* taxa from Croatia. *Biochem. Syst. Ecol.* **2006**, *34*, 875–881. [CrossRef]
2. Domac, R. *Flora of Croatia and Neighbouring Areas*; Školska Knjiga: Zagreb, Hrvatska, 1973; pp. 332–335.
3. Ball, P.W. *Stachys* L. In *Flora Europaea*; Tutin, T.G., Heywood, V.H., Burges, N.A., Moore, D.M., Valentine, D.H., Walters, S.M., Webb, D.A., Eds.; Cambridge University Press: Cambridge, UK, 1972; Volume 3, pp. 151–157.
4. Bhattacharjee, R. Taxonomic studies in *Stachys*: II. A new infrageneric classification of *Stachys* L. *Notes Roy. Bot. Gard. Edinbourgh* **1980**, *38*, 65–96.
5. Mulligan, G.A.; Munro, D.B. Taxonomy of species of North American *Stachys* (Labiatae) found in north of Mexico. *Rev. Ecol. Syst.* **1989**, *116*, 35–51.
6. Turner, B.L. Synopsis of Mexican and Central American species of *Stachys* (Lamiaceae). *Phytologia* **1994**, *77*, 338–377. [CrossRef]
7. Falciani, L. Systematic revision of *Stachys* sect. *Eriostomum* (Hofmans & Link) Dumort. in Italy. *Lagascalia* **1997**, *19*, 187–237.
8. Lovašen-Ebberhardt, Ž. *Stachys* L. In *Index Florae Croaticae 3*; Nikolić, T., Ed.; Hrvatski Prirodoslovni Muzej: Zagreb, Croatia, 2000; pp. 24–25.
9. Šilić, Č.; Šolić, M.E. The taxonomy, chorology and ecology of *Stachys menthifolia* Vis. (Lamiaceae) in the north-west part of its distribution area. *Acta Bot. Croat.* **2002**, *61*, 51–56.
10. Bilušić Vundać, V. Pharmacobotanical and Chemotaxonomical Characterisation of Some *Stachys* Taxa (Lamiaceae). Ph.D. Thesis, University of Pharmacy and Biochemistry, Zagreb, Croatia, 2006.
11. Forenbacher, S. *Velebit i Njegov Biljni Svijet*; Školska knjiga: Zagreb, Croatia, 1990; pp. 581–583.
12. Region Bojňanský, V.; Fargašová, A. *Atlas of Seeds and Fruits of Central and East-European Flora The Carpathian Mountains*; Springer: Dordrecht, The Netherlands, 2007; p. 1046. ISBN1 978-1-4020-5361-0. ISBN2 978-1-4020-5362-7.
13. Visiani, R. Plantae Rariores in Dalmatia Recens Detectae. *Flora (Regensb.)* **1829**, *12*, 1–24.
14. Ćavar, S.; Maksimović, M.; Šolić, M.E. Comparison of Essential Oil Composition of *Stachys menthifolia* Vis. from Two Natural Habitats in Croatia. *Biol. Nyssana* **2010**, *1*, 99–103.

15. Ćavar, S.; Maksimović, M.; Vidic, D.; Šolić, M.E. Chemical composition of the essential oil of *Stachys menthifolia* Vis. *Pharm. Biol.* **2010**, *48*, 170–176. [CrossRef]
16. Chalchat, J.C.; Petrović, S.D.; Maksimović, Z.A.; Gorunović, M.S. Essential oil of *Stachys officinalis* (L.) Trevis, Lamiaceae, from Montenegro. *J. Essent. Oil Res.* **2001**, *13*, 286–287. [CrossRef]
17. Grujić-Jovanović, S.; Skaltsa, H.D.; Marin, P.; Soković, M. Composition and antibacterial activity of the essential oil of six *Stachys* species from Serbia. *Flavour Fragr. J.* **2004**, *19*, 134–144. [CrossRef]
18. Lazarević, J.S.; Đorđević, A.S.; Kitić, D.V.; Zlatković, B.K.; Stojanović, G.S. Chemical composition and antimicrobial activity of the essential oil of *Stachys officinalis* (L.) Trevis. (Lamiaceae). *Chem. Biodivers.* **2013**, *10*, 1335–1349. [CrossRef]
19. Tirillini, B.; Pellegrino, R.; Bini, L.M. Essential oil composition of *Stachys sylvatica* from Italy. *Flavour Fragr. J.* **2004**, *19*, 330–332. [CrossRef]
20. Hajdari, A.; Novak, J.; Mustafa, B.; Franz, C. Essential oil composition and antioxidant activity of *Stachys sylvatica* L. (*Lamiaceae*) from different wild populations in Kosovo. *Nat. Prod. Res.* **2012**, *26*, 1676–1681. [CrossRef] [PubMed]
21. Kiliç, Ö.; Özdemir, F.; Şinasi, Y. Essential oils and fatty acids of *Stachys* L. taxa, a chemotaxonomic approach. *Prog. Nutr.* **2017**, *19*, 49–59. [CrossRef]
22. Renda, G.; Bektaş, N.Y.; Korkmaz, B.; Celik, G.; Sevgi, S.; Yayl, N. Volatile Constituents of three *Stachys* L. species from Turkey. *Marmara Pharm. J.* **2017**, *21*, 278–285. [CrossRef]
23. Senatore, F.; Formisano, C.; Rigano, D.; Piozzi, F.; Rosselli, S. Chemical composition of the essential oil from aerial parts of *Stachys palustris* L. (Lamiaceae) growing wild in Southern Italy. *Croat. Chem. Acta* **2007**, *80*, 135–139.
24. Conforti, F.; Menichini, F.; Formisano, C.; Rigano, D.; Senatore, F.; Arnold, N.A.; Piozzi, F. Comparative chemical composition, free radical-scavenging and cytotoxic properties of essential oils of six *Stachys* species from different regions of the Mediterranean Area. *Food Chem.* **2006**, *116*, 898–905. [CrossRef]
25. Skaltsa, H.D.; Demetzos, C.; Lazari, D.; Sokovic, M. Essential oil analysis and antimicrobial activity of eight *Stachys* species from Greece. *Phytochemistry* **2003**, *64*, 743–752. [CrossRef]
26. Goren, A.C.; Piozzi, F.; Akçicek, E.; Kılıc, T.; Carıkc, S.; Mozioğlu, E.; Setzer, W.N. Essential oil composition of twenty-two *Stachys* species (mountain tea) and their biological activities. *Phytochem. Lett.* **2011**, *4*, 448–453. [CrossRef]
27. Demirci, B.; Yıldız, G.; Kırımer, N.; Ocak, A.; Hüsnü Can Başer, K. Essential oil composition of *Stachys obliqua* Waldst. et Kit. *Nat. Volatiles Essent. Oils* **2018**, *5*, 17–22.
28. Harmandar, M.; Duru, M.E.; Cakir, A.; Hirata, T.; Izumi, S. Volatile constituents of *Stachys obliqua* L. (Lamiaceae) from Turkey. *Flavour Fragr.* **1997**, *12*, 211–213. [CrossRef]
29. Maly, E. Paper chromatography of the essential oils occuring in the genus *Stachys*. *J. Chromatogr.* **1985**, *333*, 288–289. [CrossRef]
30. Chalchat, J.C.; Petrović, S.D.; Maksimović, Z.A.; Gorunović, M.S. Essential oil of the herb of *Stachys recta* L., Lamiaceae, from Serbia. *J. Essent. Oil Res.* **2000**, *12*, 455–458. [CrossRef]
31. Cakir, A.; Duru, M.E.; Harmandar, M.; Izumi, S.; Hirata, T. The volatile constituents of *Stachys recta* L. and *Stachys balansae* L. from Turkey. *Flavour Fragr. J.* **1997**, *12*, 215–218. [CrossRef]
32. Bilušić Vundać, V.; Brantner, AH.; Miško, P. Content of polyphenolic constituents and antioxidant activity of some *Stachys* taxa. *Food Chem.* **2007**, *104*, 1277–1281. [CrossRef]
33. Bilušić Vundać, V.; Maleš, Ž.; Plazibat, M.; Golja, P.; Cetina-Čižmek, B. HPTLC determination of flavonoids and phenolic acids in some Croatian *Stachys* taxa. *J. Planar Chromatogr.* **2005**, *18*, 269–273. [CrossRef]
34. Karioti, A.; Bolognesi, L.; Vincieri, F.F.; Bilia, A.R. Analysis of the constituents of aqueous preparations of *Stachys recta* by HPLC–DAD and HPLC–ESI-MS. *J. Pharm. Biomed. Anal.* **2010**, *53*, 15–23. [CrossRef]
35. Tundis, R.; Peruzzi, L.; Menichini, F. Phytochemical and biological studies of *Stachys* species in relation to chemotaxonomy: A review. *Phytochemistry* **2014**, *102*, 7–39. [CrossRef] [PubMed]
36. Frezza, C.; Venditti, A.; Maggi, F.; Cianfaglione, K.; Nagy, D.U.; Serafini, M.; Bianco, A. Secondary metabolites from *Stachys palustris* L. In *Proceedings of the 6th International Congress of Aromatic and Medicinal Plants (CIPAM 2016), Coimbra, Portugal, 29 May–1 June 2016*; Salgueiro, L., Cavaleiro, C., Cabral, C., Eds.; Universidade de Coimbra: Coimbra, Portugal, 2016; p. 80. ISBN 978-989-95050-1-8.

37. Bilušić Vundać, V. Qualitative and quantitative determination of amino acids in some Stachys taxa—Part of PhD thesis 'Pharmacobotanical and chemotaxonomical characterisation of some Stachys taxa (Lamiaceae). 2006. Available online: https://www.researchgate.net/publication/327651956_Qualitative_and_quantitative_determination_of_aminoacids_in_some_Stachys_taxa_part_of_PhD_thesis_%27Pharmacobotanical_and_chemotaxonomical_characterisation_of_Some_Stachys_taxa_Lamiaceae%27 (accessed on 2 December 2018).
38. Chowdhury, T.; Mandal, A.; Roy, SC.; De Sarker, D. Diversity of the genus *Ocimum* (Lamiaceae) through morpho-molecular (RAPD) and chemical (GC–MS) analysis. *J. Genet. Eng. Biotechnol.* **2017**, *15*, 275–286. [CrossRef]

© 2019 by the author. Licensee MDPI, Basel, Switzerland. This article is an open access article distributed under the terms and conditions of the Creative Commons Attribution (CC BY) license (http://creativecommons.org/licenses/by/4.0/).

Article

On the Possible Chemical Justification of the Ethnobotanical Use of *Hyptis obtusiflora* in Amazonian Ecuador

Carmen X. Luzuriaga-Quichimbo [1], José Blanco-Salas [2,*], Carlos E. Cerón-Martínez [3], Milan S. Stanković [4] and Trinidad Ruiz-Téllez [2]

1. Centro de Investigación Biomédica, Facultad de Ciencias de la Salud Eugenio Espejo, Universidad Tecnológica Equinoccial, Av. Mariscal Sucre y Mariana de Jesús, Quito 170527, Ecuador; luzuriaga.cx@gmail.com
2. Department of Vegetal Biology, Ecology and Earth Science, Faculty of Sciences, University of Extremadura, 06071 Badajoz, Spain; truiz@unex.es
3. Herbario Alfredo Paredes, QAP, Universidad Central de Ecuador, Quito 170129, Ecuador; cecm57@yahoo.com
4. Department of Biology and Ecology, Faculty of Science, University of Kragujevac, 34000 Kragujevac, Serbia; mstankovic@kg.ac.rs
* Correspondence: blanco_salas@unex.es; Tel.: +34-924-289-300

Received: 15 September 2018; Accepted: 20 November 2018; Published: 23 November 2018

Abstract: In rural areas of Latin America, *Hyptis* infusions are very popular. *Hyptis obtusiflora* extends from Mexico throughout Central America to Bolivia and Peru. It has added value in Ecuador where it has been used by different ethnic groups. We aimed to learn about the traditional knowledge of ancient Kichwa cultures about this plant, and to contrast this knowledge with the published information organized in occidental databases. We proposed to use traditional knowledge as a source of innovation for social development. Our specific objectives were to catalogue the uses of *H. obtusiflora* in the community, to prospect on the bibliography on a possible chemical justification for its medicinal use, to propose new products for development, and to give arguments for biodiversity conservation. An ethnobotanical survey was made and a Prisma 2009 Flow Diagram was then followed for scientific validation. We rescued data that are novel contributions for the ethnobotany at the national level. The catalogued main activity of anti-inflammation can be related to the terpene composition and the inhibition of xanthine oxidase. This opens the possibility of researching the extract of this plant as an alternative to allopurinol or uricosuric drugs. This is a concrete example of an argument for biodiversity conservation.

Keywords: *Hyptis*; Lamiaceae; Kichwa; terpene; caryophyllene; xanthine oxidase; gut; anti-inflammatory; uricosuric

1. Introduction

The genus *Hyptis* is the second most important of the American Lamiaceae. It is made up of at least 290 species which are almost exclusively Neotropical [1,2]. Plants of this genus are covered with glandular trichomes that produce essential oils that make them very popular in rural areas of Latin America where they are taken as an infusion to treat respiratory and gastrointestinal disorders or skin diseases [3,4]. Different pharmacological activities (antibacterial [5,6] anti-inflammatory [7], antiparasitic [8], and antiproliferative [9]) have been tested experimentally in this genus, so it is an interesting taxon in the field of medical applications. Some *Hyptis* have been used against malaria (*Hyptis mutabilis* (Rich.) Briq. [10]), and others have been used against scabies (*Hyptis suaveolens* (L.) Poit. [11]) or as insect repellents (*Hyptis tafallae* Benth [12]). Nearly 20 species have been reviewed from the phytochemical point of view [13]. The volatile oil is rich in monoterpenes and sesquiterpenes and have different compositions characterized by the occurrence of major components such as E-caryophyllene, 1,8-cineole, and sabinene or others such as eugenol or cadinene. Minor constituents, which are almost always present, are *p*-cymene and α-pinene [13].

Hyptis is the best assorted genus of the genuine, useful Ecuadorian Lamiaceae. In Ecuador, the majority of Lamiaceae plants with catalogued uses are allochthonous. Contrary to this, there are 10 autochthonous *Hyptis* species included in the list of useful plants from Ecuador [14].

They have interesting chemical compounds and biological activities as reported by experimental works published in the scientific literature (Table 1), but their traditional knowledge is mostly reduced to small ethnic groups and areas, and are either endangered or near extinct [14].

Table 1. Biological activity and chemical components identified (bibliographic references in brackets) for *Hyptis* species from the Catalogue of Useful Plants of Ecuador.

Species	Principal Component		Activities	References
H. atrorubens Poit.	E-caryophyllene	sesquiterpene	Anti-inflammatory	[15]
H. capitata Jacq.	eugenol	monoterpene	antimicrobial	[16]
H. eriocephala Benth.	nepetoidine	sesquiterpene	antifungal, antifeedant	[17]
H. mutabilis Briq.	E-caryophyllene	sesquiterpene	Anti-inflammatory	[18]
H. obtusata Benth.	--	---	ns	[19]
H. pectinata	pectinolide	furanone	antinociceptive	[20,21]
H. purdiei Benth.	--	--	ns	[22]
H. recurvata Poit.	1,8 cineol	monoterpene	Anti-inflammatory	[23]
H. verticillata Jacq.	17 lignans	others	antineoplasic	[24]

The species with a more widespread use in the Ecuadorian catalogue is *Hyptis obtusiflora* (Figure 1), perhaps because of a certain colonizing character and a facility to occupy habitats of tropical rain forests. In Ecuador, it has been collected from several locations of the Coast and Sierra Regions and in the northern-most Amazonian provinces such as Orellana or Napo. It can be found from sea level to the montane (0–1800 m). Its natural distribution area extends north from Peru (where it was first described [25]) throughout Central America to Mexico [26]. In Peru, the national inventory of the Ministry on Traditional Medicine picks up its use against ringworms, head wounds, and pharyngitis [27], which comes from the Amazonic Yanesha populations who have an ancestral knowledge of the plant they called "ollamepan pasheñorrer". There is very little registered bibliography from other countries.

Figure 1. *Hyptis obtusiflora* [28].

In the frame of the Convention on Biological Diversity and the corresponding procedures of the Nagoya Protocol [29,30], we have considered engaging in the contribution of arguments for biodiversity conservation in Amazonian Ecuador in order to deepen the knowledge of and research on *Hyptis obtusiflora*. We have selected the scarcely contacted habitats of River Bobonaza to learn about the traditional knowledge of ancient Kichwa cultures about this plant and to justify the applications of the species through a critical bibliographic prospection from the experimental sciences perspective. We propose to use traditional knowledge as a source of innovation for social development.

For all these reasons, we proposed, as specific objectives of this paper:

1. To catalogue some traditional and threatened uses of *H. obtusiflora*,
2. To justify the reasons for its consumption by bibliographic prospection,
3. To propose new research or products for development, and
4. To give arguments for biodiversity conservation.

2. Results and Discussion

- *On the Singularity of the Studied Plant and the Value of the Surveyed Ethnobotanical Community*

The conservation of traditional knowledge allows us to report on real testimonies of little-contacted populations located in the Amazon, whose life has developed around plant biodiversity and the use of floristic resources. However, this work fulfills one of the goals of the Ecuador National Biodiversity Research Agenda dealing with prioritizing research in those gaps of knowledge on Ecuador's biodiversity [31].

Hyptis obtusiflora ("mule shell" or "secret of Indian" in Spanish, "taku taku" in Awapit or "waka muké" in Shuar-Chicham) had not been previously collected nor cited in the Bobonaza Basin, which occupies most of Pastaza, the biggest province of the country (29.641 km^2); this paper is a first advance. Regardless of this, in Ecuador, the Mestizo population uses its juice to heal wounds, and in Pichincha province, infusions are made for hot baths; the Awa from Carchi bath their legs with the ashes made from the burning the plant; the Shuar of Napo and Orellana cook the leaves to relieve the flu and to fight skin infections; and the Chachis of Esmeraldas prepare drinks with macerated leaves to relieve stomach pain [14].

Our field work revealed that the Kichwa of the Bobonaza in Pakayaku used *Hyptis obtusiflora* externally as a medicinal plant on the skin to treat stings, pimples, or injuries that insects cause, especially in the most vulnerable individuals of the population, i.e., the children. They use the juice of the leaf, and the guideline is a twice-a-day application. The people of the community value this remedy very positively. The transmission of this knowledge has occurred orally in the ayllus (families) through specific conversations among women. It is a current use and very often used at the frequency with which these injuries occur in the location where the climate promotes an abundance of insects. The plant is consumed just after collection; it has never been stored or preserved.

The data we collected had not been documented among the Kichwas ethnic group in the country. These aspects and the vernacular name given in the community (karacha panga) in the Kichwa language are novel contributions for ethnobotany at the national level. The conservation and sustainable use of biological diversity, particularly of little-known species in vulnerable ecosystems such as the tropical forests of these communities of Pastaza, which remains nowadays underexplored from the botanical point of view [32], is a worthwhile task.

- *Towards an Explanation of the Use of the Plant: Xanthine Oxidase (XO) Inhibition*

Xanthine oxidase (XO) activates inflammatory processes [33] by stimulating the production of reactive oxygen species [34]. The metanolic extract (1 µg/mL) of this plant has been demonstrated to inhibit 40% of XO enzymatic activities [35]. The extract of *H. obtusiflora* can be considered anti-inflammatory for this reason.

The specific chemical profile of *Hyptis obtusiflora* has not yet been studied, although there is information about co-generic species and data on their chemical profiles (see Table 1). The biological applications of some of the components isolated from close species are well known as anti-inflammatories or antinociceptives in several cases. Thus, 1,8-cineole has shown therapeutic benefits in inflammatory diseases, such as asthma and chronic obstructive pulmonary disease [36]; E-caryophyllene is an antiinflammatory molecule because (a) it produces an anti-spasmodic activity on smooth muscle, which is probably related to Ca^{2+} channel blockade [37]; and (b) it induces inhibition of cytokines involved in the arachidonic acid and histamine pathways [38].

Finally, we must emphasize how the inhibitory activity of XO that was firstly tested at biochemical laboratories [35] has recently been tested in silico using docking software [34]. This virtual approach has revealed that the inhibitory enzymatic action is related to the structural affinity of XO and E-caryophyllene [34]. This is consistent with the foregoing [38] experimental publications.

- *Innovative and Clinical Possibilities for H. obtusiflora Use and Conservation under the Nagoya Protocol*

Innovative drug discovery in anti-inflammatories today is focused in non-purine structures which offer some advantages over existing commercialized molecules such as allopurinol. To date, the alternative drug had been febuxostat, but the USA (Food and Drug Administration) FDA recently issued safety alerts for its use [39]. This research speciality is oriented to treat hyperuricemia, gout, ulcers, cancer, ischemia, hypertension, and oxidative damage [40]. For this reason, it seems very interesting to carry out new research lines focused on the characterization of the chemical profile of the essential oil and the chemotypes of *Hyptis obtusiflora* in the Kichwa communities of the Bobonaza River Basin in particular and the Amazonian Ecuador in general. It is a very useful plant to be cultivated because it is an autochthonous element with an easy method of propagation, and the essential oil can give many medicinal benefits. The pharmacological activity of the components dealing with XO inhibition and anti-inflammatory processes must be quantified and standardized. This is a good case for demonstrating the possibilities that vegetal biodiversity can offer as a resource to improve human health and quality of life. Innovative products, to substitute classical lifelong anti-gut drugs (e.g., allopurinol or, more recently, flebuxostat) can be developed from the jungle biodiversity.

The clinical possibilities of the extract from this plant for anti-inflammatory pharmacology seem promising, although a precise chemical profile characterization is still needed. This opens several lines of phytochemical investigation that can ratify the interest for its conservation. In this case, the Nagoya Protocol cannot be forgotten and the Pakayaku Kichwa Original Community must be taken into account.

3. Materials and Methods

3.1. Ethnobotanical Survey

The selected community lies in a fairly isolated region where bio- and ethnodiversity studies are still lacking (Bobonaza River, Pastaza, Ecuador). It is named Pakayaku. One of us (C.X.L.-Q.) was allowed to visit due to the environmental and education programs she has been conducting since 2008 from the Biological Station Pindo Mirador in the northern Bobonaza River Basin (S 1°27'09''–W 78°04'51'').

Plant collection permits were granted by the Ministry of the Environment. Plant vouchers were deposited at the Herbarium José Alfredo Paredes, Universidad Central de Ecuador, Quito QAP Herbarium: Ecuador, Pastaza: Sarayaku, Pakayaku, banks of the Bobonaza River, sector of Chumbi yaku, path to chacra Sra. Ana Aranda, 402 m, 01°39'36.4'' S, 077°36'55.4'' W, lowland evergreen forest, 2 October 2015, *C. X. Luzuriaga-Q & H. Manya* (QAP 93168). Identification was revised by C. Cerón.

The interviews and ethnobotanical protocols were as described previously [41].

Collective written research consent was granted by Ms. Luzmila Gayas, the community president of the Assembly of Pakayaku. Prior oral individual consent was obtained from the persons taking part in our survey. Planned house visits and walking routes accompanied by Kichwa interpreters and local inhabitants of Pakayaku were made. Interviews were semi-structured and included a series of open questions aimed to encourage discussion. All interviews were recorded. Four knowledgeable elders of the Pakayaku community acted as informants and agreed to reveal their wisdom of the karacha panga. The informants answered freely about several topics, namely the Kichwa common name, the parts of the plant used, the description of use, the harvest season, storage (if any), concoction, and treatment targets. After the field wok, data were added to a Microsoft Excel spreadsheet. All recorded uses were referred to a previously published classification [14]. The data provided by the community were compared with the existing ethnobotanical literature from Ecuador [14].

3.2. Bibliography Review and Justification of the Activity

A bibliographic study was performed to provide scientific evidence for the medicinal uses of the plant. The accessed databases were: ISI, Scopus, Dialnet, SpaceNet, MEDLINE, PubMed, ScienceDirect, Google Patents and Scholar, and Wiley Online. The methodological Prisma 2009 Flow Diagram [42] was followed.

Author Contributions: Conceptualization, T.R.-T.; methodology, C.E.C.-M. and C.X.L.-Q.; validation, J.B.-S.; formal analysis, M.S.; investigation, C.X.L.-Q.; data curation, C.E.C.-M. and C.X.L.-Q.; writing-original draft preparation, T.R.-T.; writing-review and editing, J.B.-S.; visualization, M.S.S.; supervision, M.S.S.; project administration, T.R.-T.; funding acquisition, C.X.L.-Q. and T.R.-T.

Funding: This research was partially funded by the Government of Extremadura (Spain) and the European Union through the action Apoyos a los Planes de Actuación de los Grupos de Investigación Catalogados de la Junta de Extremadura: FEDER GR15080.

Acknowledgments: We are grateful to the members of the Kichwa community of Pakayaku, Luzmila Gayas, the People's Assembly of Pakayaku, and the collaborating ayllus (families), for their cooperation during the field work. M.V. Gil Alvarez (Organic Chemistry Department, University of Extremadura) assisted us with the chemical drawing software.

Conflicts of Interest: The authors declare no conflict of interest. The funding sponsors had no role in the design of the study; in the collection, analyses, or interpretation of data; in the writing of the manuscript; or in the decision to publish the results.

References

1. Kubitzki, K.; Kadereit, J.W. Flowering plants • Dicotyledons. In *The Families and Genera of Vascular Plants*; Springer: Berlin, Germany, 2004.
2. Harley, R.M.; Pastore, J.F.B. A generic revision and new combinations in the *Hyptidinae* (Lamiaceae), based on molecular and morphological evidence. *Phytotaxa* **2012**, *58*, 1–55. [CrossRef]
3. Arruda, R.C.O.; Araujo, C.; Farias, C.S.; Victório, C.P.; Pott, V.J. Contribución de la epidermis en la identificación taxonómica de *Hyptis* (Lamiaceae) nativo de Brasil. V Jornadas Nacionales de Plantas Aromáticas Nativas y sus aceites esenciales. I Jornadas Nacionales de Plantas Medicinales Nativas. Brasil. *Dominguezia* **2016**, *32*, 75–76.
4. Pinheiro, M.A.; Magalhães, R.; Torres, D.; Cavalcante, R.; Mota, F.X.; Oliveira Coelho, E.A.; Moreira, H.; Lima, G.; da Costa Araújo, P.; Cardoso, J.L.; et al. Gastroprotective effect of alpha-pinene and its correlation with antiulcerogenic activity of essential oils obtained from *Hyptis* species. *Pharmacogn. Mag.* **2015**, *11*, 123–130. [CrossRef]
5. Andrade, A.M.; Oliveira, J.P.R.; Santos, A.L.L.M.; Franco, C.R.P.; Antoniolli, Â.R.; Estevam, C.S.; Thomazzi, S.M. Preliminary study on the anti-inflammatory and antioxidant activities of the leave extract of *Hyptis fruticosa* Salzm. ex Benth., Lamiaceae. *Rev. Bras. Farmacogn.* **2010**, *20*, 962–968. [CrossRef]
6. Tesch, N.R.; Yánez, R.M.; Rojas, X.M.; Rojas-Fermín, L.; Carrillo, J.V.; Díaz, T.; Vivas, F.M.; Colmenares, C.Y.; González, P.M. Composición química y actividad antibacteriana del aceite esencial de *Hyptis suaveolens* (L.) Poit. (Lamiaceae) de los Llanos venezolanos. *Rev. Peru. Biol.* **2015**, *22*, 103–107. [CrossRef]

7. Simoes, R.R.; Coelho, I.D.; Junqueira, S.C.; Pigatto, G.R.; Salvador, M.J.; Santos, A.R.; de Faria, F.M. Oral treatment with essential oil of *Hyptis spicigera* Lam. (Lamiaceae) reduces acute pain and inflammation in mice: Potential interactions with transient receptor potential (TRP) Ion Channels. *J. Ethnopharmacol.* **2017**, *200*, 8–15. [CrossRef] [PubMed]
8. Valadeau, C.; Pabon, A.; Deharo, E.; Albán Castillo, J.; Esteve, Y.; Augusto, L.F.; Rojas, R.; Gamboa, D.; Sauvain, M.; Castillo, D.; et al. Medicinal Plants from the Yanesha (Peru): Evaluation of the leishmanicidal and antimalarial activity of selected extracts. *J. Ethnopharmacol.* **2009**, *123*, 413–422. [CrossRef] [PubMed]
9. Taylor, P.; Arsenak, M.; Abad, M.J.; Fernandez, A.; Milano, B.; Gonto, R.; Ruiz, M.-C.; Fraile, S.; Taylor, S.; Estrada, O.; et al. Screening of venezuelan medicinal plant extracts for cytostatic and cytotoxic activity against tumor cell lines. *Phytother. Res.* **2013**, *27*, 530–539. [CrossRef] [PubMed]
10. García, H. *Flora Medicinal de Colombia*; Instituto de Ciencias Naturales de la Universidad Nacional de Bogotá: Bogotá, Colombia, 1975.
11. Sharma Prince, P.; Roy Ram, K.; Anurag, G.; Sharma, V. *Hyptis suaveolens* (L.) Poit: A phyto-pharmacological review. *Int. J. Chem. Pharm. Sci.* **2013**, *4*, 1–11.
12. Luis Fernandez-Alonso, J.A. New species of *Hyptis* (Labiatae) from Colombia. *Anal. Jard. Bot. Madr.* **2010**, *67*, 127–135. [CrossRef]
13. McNeil, M.; Facey, P.; Porter, R. Essential oils from the *Hyptis* genus—A Review (1909–2009). *Nat. Prod. Commun.* **2011**, *6*, 1775–1796. [PubMed]
14. De la Torre, L.; Navarrete, H.; Muriel, P.; Marcia, M.; Balslev, H. *Enciclopedia De Plantas Utiles Del Ecuador*; Herbario QCA de la Escuela de Ciencias Biológicas de la Pontificia Universidad Católica del Ecuador & Herbario AAU del Departamento de Ciencias Biológicas de la Universidad de Aarhus: Quito, Ecuador, 2008.
15. Kerdudo, A.; Njoh Ellong, E.; Gonnot, V.; Boyer, L.; Michel, T.; Adenet, S.; Rochefort, K.; Fernandez, X. Essential oil composition and antimicrobial activity of *Hyptis atrorubens* Poit. from Martinique (F.W.I.). *J. Essent. Oil Res.* **2016**, *28*, 436–444. [CrossRef]
16. Rupa, D.; Sulistyaningsih, Y.C.; Dorly, D.; Ratnadewi, D. Identification of secretory structure, histochemistry and phytochemical compounds of medicinal plant *Hyptis capitata* Jacq. *J. Biotropia* **2017**, *24*, 94–103. [CrossRef]
17. Grayer, R.J.; Eckert, M.R.; Veitch, N.C.; Kite, G.C.; Marin, P.D.; Kokubun, T.; Simmonds, M.S.J.; Paton, A.J. The chemotaxonomic significance of two bioactive caffeic acid esters, nepetoidins a and b, in the Lamiaceae. *Phytochemistry* **2003**, *64*, 519–528. [CrossRef]
18. Aguiar, E.H.A.; Zoghbi, M.D.G.B.; Silva, M.H.L.; Maia, J.G.S.; Amasifén, J.M.R.; Rojas, U.M. Chemical Variation in the essential oils of *Hyptis mutabilis* (Rich.) Briq. *J. Essent. Oil Res.* **2003**, *15*, 130–132. [CrossRef]
19. Florence, A.; Joselin, J.; Sukumaran, S.; Jeeva, S. Screening of phytochemical constituents from certian flower extracts. *Int. J. Pharm. Rev. Res.* **2014**, *4*, 152–159.
20. Paixão, M.S.; Melo, M.S.; Oliveira, M.G.B.; Santana, M.T.; Lima, A.C.B.; Damascena, N.P.; Dias, A.S.; Araujo, B.S.; Estevam, C.S.; Botelho, M.A.; et al. *Hyptis pectinata*: Redox protection and orofacial antinociception. *Phyther. Res.* **2013**, *27*, 1328–1333. [CrossRef] [PubMed]
21. Boalino, D.M.; Connolly, J.D.; McLean, S.; Reynolds, W.F.; Tinto, W.F. α-pyrones and a 2(5h)-furanone from *Hyptis pectinata*. *Phytochemistry* **2003**, *64*, 1303–1307. [CrossRef] [PubMed]
22. Tinitana, F.; Rios, M.; Romero-Benavides, J.C.; de la Cruz Rot, M.; Pardo-de-Santayana, M. Medicinal plants sold at traditional markets in southern Ecuador. *J. Ethnobiol. Ethnomed.* **2016**, *12*. [CrossRef] [PubMed]
23. Leclercq, P.A.; Delgado, H.S.; Garcia, J.; Hidalgo, J.E.; Cerruttti, T.; Mestanza, M.; Ríos, F.; Nina, E.; Nonato, L.; Alvarado, R.; et al. Aromatic plant oils of the Peruvian Amazon. Part 2. *Cymbopogon citratus* (DC) Stapf., *Renealmia* Sp., *Hyptis recurvata* Poit. and *Tynanthus panurensis* (Bur.) Sandw. *J. Essent. Oil Res.* **2000**, *12*, 14–18. [CrossRef]
24. Picking, D.; Delgoda, R.; Boulogne, I.; Mitchell, S. *Hyptis verticillata* Jacq: A review of its traditional uses, phytochemistry, pharmacology and toxicology. *J. Ethnopharmacol.* **2013**, *147*, 16–41. [CrossRef] [PubMed]
25. Bentham, G. *Labiatarum Genera et Species Fasc.2. VI*; James Ridgway and Sons: London, UK, 1833.
26. Missouri Botanical Garden. Tropicos Database. 2018. Available online: http://www.tropicos.org (accessed on 1 September 2017).
27. Ministerio de Salud de Perú. Inventario Nacional de Plantas Medicinales. Available online: http://www.portal.ins.gob.pe/es/censi/censi-c4/plantas-medicinales/inventario-nacional-de-plantas-medicinales (accessed on 1 September 2017).

28. Botanicalillustrations.org/*Hyptis-obtusiflora*. Available online: http://botanicalillustrations.org/illustration.php?id_illustration=117540&SID=0&mobile=0&code_category_taxon=9&size=1 (accessed on 10 September 2018).
29. United-Nations. Convention on Biological Diversity 2010. Available online: https://www.cbd.int/convention/ (accessed on 1 September 2017).
30. Convention on Biological Diversity. 2010 CoP10 Decisions. Available online: https://www.cbd.int/decisions/cop (accessed on 2 April 2018).
31. *Agenda Nacional de Investigación Sobre la Biodiversidad*; MAE, Senescyt, INABIO: Quito, Ecuador, 2018.
32. Luzuriaga-Quichimbo, C.X. *Estudio Etnobotánico en Comunidades Kichwas Amazónicas de Pastaza*; Ecuador, Universidad de Extremadura: Caceres, España, 2017.
33. Grassmann, J.; Hippeli, S.; Dornisch, K.; Rohnert, U.; Beuscher, N.; Elstner, E.F. Antioxidant properties of essential oils. possible explanations for their anti-inflammatory effects. *Arzneimittelforschung* **2000**, *50*, 135–139. [PubMed]
34. Umamaheswari, M.; Prabhu, P.; Asokkumar, K.; Sivashanmugam, T.; Subhadradevi, V.; Jagannath, P.; Madeswaran, A. In silico docking studies and in vitro xanthine oxidase inhibitory activity of commercially available terpenoids. *Int. J. Phytopharm.* **2012**, *4*, 3460–3462. [CrossRef]
35. González, A.G.; Bazzocchi, I.L.; Moujir, L.; Ravelo, A.G.; Correa, M.D.; Gupta, M.P. Xanthine Oxidase inhibitory activity of some panamanian plants from Celastraceae and Lamiaceae. *J. Ethnopharmacol.* **1995**, *46*, 25–29. [CrossRef]
36. Juergens, U.R. Anti-Inflammatory properties of the monoterpene 18-cineole: Current Evidence for Co-Medication in Inflammatory Airway Diseases. *Drug Res.* **2014**, *64*, 638–646. [CrossRef]
37. Pinho-Da-Silva, L.; Mendes-Maia, P.V.; Do Nascimento Garcia Teófilo, T.M.; Barbosa, R.; Ceccatto, V.M.; Coelho-De-Souza, A.N.; Cruz, J.S.; Leal-Cardoso, J.H. Trans-caryophyllene, a natural sesquiterpene, causes tracheal smooth muscle relaxation through blockade of voltage-dependent Ca^{2+} channels. *Molecules* **2012**, *17*, 11965–11977. [CrossRef] [PubMed]
38. De Morais Oliveira-Tintino, C.D.; Pessoa, R.T.; Fernandes, M.N.M.; Alcântara, I.S.; da Silva, B.A.F.; de Oliveira, M.R.C.; Martins, A.O.B.P.B.; do Socorro da Silva, M.; Tintino, S.R.; Rodrigues, F.F.G.; et al. Anti-inflammatory and anti-edematogenic action of the Croton Campestris A. St.-Hil (Euphorbiaceae) essential oil and the compound β-caryophyllene in in vivo models. *Phytomedicine* **2018**, *41*, 85–92. [CrossRef] [PubMed]
39. White, W.B.; Chohan, S.; Dabholkar, A.; Hunt, B.; Jackson, R. Cardiovascular safety of febuxostat and allopurinol in patients with gout and cardiovascular comorbidities. *Am. Heart J.* **2012**, *164*, 14–20. [CrossRef] [PubMed]
40. Kumar, R.; Darpan; Sharma, S.; Singh, R. Xanthine Oxidase Inhibitors: A patent survey. *Expert Opin. Ther. Pat.* **2011**, *21*, 1071–1108. [CrossRef] [PubMed]
41. Luzuriaga-Quichimbo, C.X.; Ruiz-Téllez, T.; Blanco-Salas, J.; Cerón Martínez, C.E. Scientific validation of the traditional knowledge of sikta ("Tabernaemontana Sananho", Apocynaceae) in the canelo-kichwa amazonian community. *Mediterr. Bot.* **2018**, *39*, 183–191. [CrossRef]
42. Moher, D.; Liberati, A.; Tetzlaff, J.; Altman, D.G.; PRISMA Group. Preferred reporting items for systematic reviews and meta-analyses: The PRISMA Statement. *PLoS Med.* **2009**, *6*, e1000097. [CrossRef] [PubMed]

© 2018 by the authors. Licensee MDPI, Basel, Switzerland. This article is an open access article distributed under the terms and conditions of the Creative Commons Attribution (CC BY) license (http://creativecommons.org/licenses/by/4.0/).

Communication

Ethnobotanical Survey, Preliminary Physico-Chemical and Phytochemical Screening of *Salvia argentea* (L.) Used by Herbalists of the Saïda Province in Algeria

Yasmina Benabdesslem [1], Kadda Hachem [1,2,*], Khaled Kahloula [1] and Miloud Slimani [1]

[1] Laboratoire de Biotoxicologie, Pharmacognosie et Valorisation Biologique des Plantes (LBPVBP), Département de Biologie, Faculté des Sciences, Université Dr. Tahar Moulay de Saida, BP 138 cité ENNASR, Saida 20000, Algeria; benabdesslem.yasmina@univ-saida.dz (Y.B.); khaled.kahloula@univ-saida.dz (K.K.); miloud.slimani@univ-saida.dz (M.S.)

[2] Laboratoire des Productions, Valorisations Végétales et Microbiennes (LP2VM), Département de Biotechnologies Végétales, Université des Sciences et de la Technologie d'Oran Mohamed Boudiaf, B.P. 1505, El-Mn'aour, Oran 31000, Algeria

* Correspondence: kadda46@hotmail.com or hachem.kadda@univ-saida.dz; Tel.: +213-661-780-404

Received: 7 November 2017; Accepted: 1 December 2017; Published: 5 December 2017

Abstract: An ethnobotanical study was carried out in the Saïda region among herbalists to evaluate the use of *Salvia argentea* (L.), a plant species native from North Africa belonging to the Lamiaceae family. Forty-two herbalists were interviewed individually, aged between 30 and 70 years, all males, 52.38% of them having received a secondary education level and having performing their duties for more than a decade. This study showed that *Salvia argentea* is used specifically in the treatment of diseases of the respiratory system. The leaves are the most commonly used part, usually in the form of powder and exclusively administered orally. The preliminary results of the physicochemical characterization and the phytochemical screening of the powdered leaves of *Salvia argentea* attest to their safety and confer them a guarantee of phytotherapeutic quality.

Keywords: *Salvia argentea* (L.); ethnobotanical servey; Saïda province; leaf powder; physico-chemical; phytochemical screening

1. Introduction

The Algerian flora in general and the region of Saïda in particular, benefit from an important reserve of plants with aromatic and medicinal characteristics. Thus, medicinal plants occupy an important place in the Algerian pharmacopoeia. Even today, they play a decisive role in the treatment of certain pathologies. Despite being one of the most impressive reserves of plants throughout the world, only 10% have been studied for their pharmacological properties [1].

The region of Saïda, by its geographical location, offers a rich and diverse vegetation. Many aromatic and medicinal plants grow there spontaneously. Interest in these plants has grown steadily in recent years.

Among these numerous medicinal plants, our study focused on *Salvia argentea* belonging to the Lamiaceae family which is a plant species originating to the Mediterranean region, in northwest Africa (northern Algeria, Morocco, and Tunisia), southern Europe (Spain, Portugal, South Italy, Sicily, Malta, Albania, Bulgaria, Slovenia, Croatia, Bosnia, Kosovo, Montenegro, Serbia, Macedonia, and Greece) and Western Asia (Turkey) [2]. This family is known for its richness in numerous chemical substances capable of demonstrating various remarkable pharmacological activities [3]. Among these substances, we mention the essential oils which mainly consist of oxygenated sesquiterpenes [2,4] and which are endowed with important biological properties, such as antimicrobial [5,6] and antioxidant activities [7].

The leaves of *Salvia argentea*, also commonly known as "Ferrache en neda", are heavily covered with a silvery down, hence its name; the leaves are soft. This plant possesses sticky stems and inflorescences. Sterile upper verticillasters consist only of bracts. White-pink flowers possess corollas three times longer than the calyx [8]. This plant differs from *Salvia patula*, by its leaves which are not heart-shaped at the base; by the upper lip of the calyx with less unequal and more distant teeth; and by its connective, more strongly toothed to the point where it widens [9]. The roots of *Salvia argentea* are thick and tuberous, which makes them resistant to heat and drought, but sensitive to humidity during winter [10].

We carried out an ethno botanical survey with herbalists working with medicinal plants. The results of this survey will allow us to identify the potential roles of *Salvia argentea* in the traditional pharmacopoeia and its effects in prophylaxis. The survey was followed by a physico-chemical and phytochemical study to detect the presence of groups of chemical families in a drug preparation based on *Salvia argentea*.

2. Results and Discussion

2.1. Ethno Botanical Survey

The ethno botanical and ethno-pharmacological investigation carried out as part of this work aimed to promote the expertise of herbalists in the Saïda region and to seek out their knowledge and know-how with medicinal plants, particularly *Salvia argentea*.

2.1.1. Frequency of Use of *Salvia argentea* According to Herbalist Profile

This practice remains very important in the Saïda zone as evidenced by the number of herbalists surveyed (42), aged between 30 and 70 years, who practice their trade either in town or in the countryside. Herbalists are male (100%), most of them are married (78.57%) and a majority had received a secondary level education (52.38%) (Figure 1). All herbalists have been working for at least a decade, which sheds light on the accumulated experience and originality of knowledge about the use of *Salvia argentea*. They all have expressed the wish to follow continuous training either nationally or internationally and to develop collaboration with modern medicine through their participation in congresses and seminars.

2.1.2. Type of Collectors of *Salvia argentea*

Interviewed herbalists use several types of collectors (Figure 2): farmers (40.48%), sedentary people (33.33%), nomads (19.05%) and shepherds (7.14%).

2.1.3. Use of *Salvia argentea* and Diseases Treated

All of the herbalists revealed that *Salvia argentea* is used in traditional medicine for therapeutic use, for the treatment of respiratory diseases. In the past, *Salvia argentea* leaves have been used against wounds, probably as a hemostatic [11], but no scientific validation has been reported so far.

2.1.4. Opinion on the Efficacy of *Salvia argentea* in the Treatment of Respiratory Diseases

The opinion of herbalists on the efficacy of *Salvia argentea* against respiratory diseases shows that 92.86% think that the traditional uses of this plant lead to a cure, while 7.14% say that the use of this plant has a relief effect only (Figure 3).

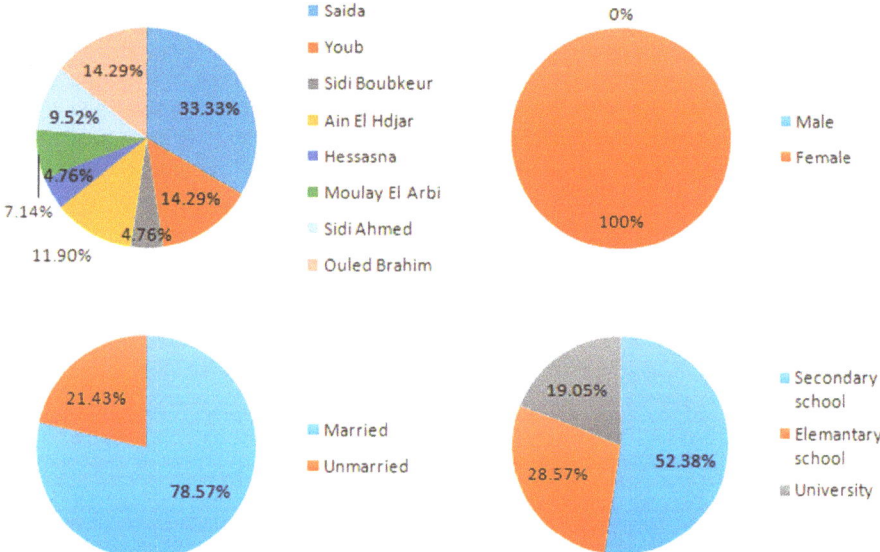

Figure 1. Information on surveyed herbalists.

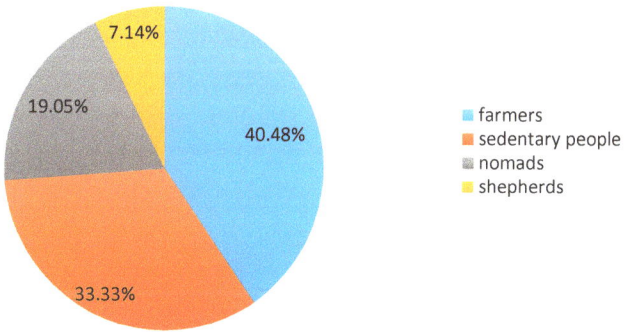

Figure 2. Type of *Salvia argentea* collectors.

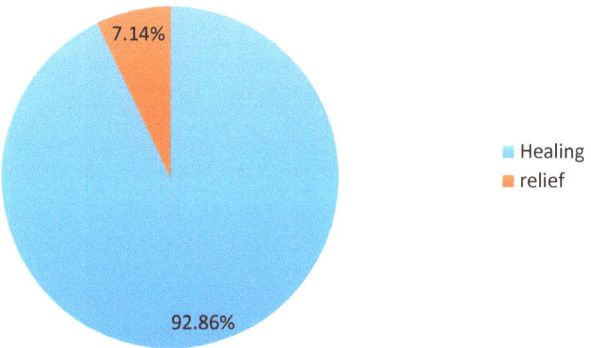

Figure 3. The opinion of herbalists on the efficacy of *Salvia argentea*.

2.1.5. The Part Used

The survey revealed that leaves are most commonly used for respiratory disease treatment with 69.05% (Figure 4), followed by roots (21.43%), and the whole plant (9.52%); however, there was no mention of any use for the inflorescences. This result is close to that established by Rhattas et al. who indicated that mainly the leaves of medicinal plants were used with a percentage of 71.75% [12]. This can be explained by the fact that leaves can be quickly harvested and that they are easy to use [13]; in addition, leaves are the main place of photosynthesis and the site of storage of many bioactive substances responsible for various biological properties [14]. The use of leaves is harmless for the regeneration of the plants and ensures the preservation of floristic richness [15]. Indeed, there is a clear relationship between the part of the plant which is exploited and the consequences of this exploitation on the persistence of this plant species [16].

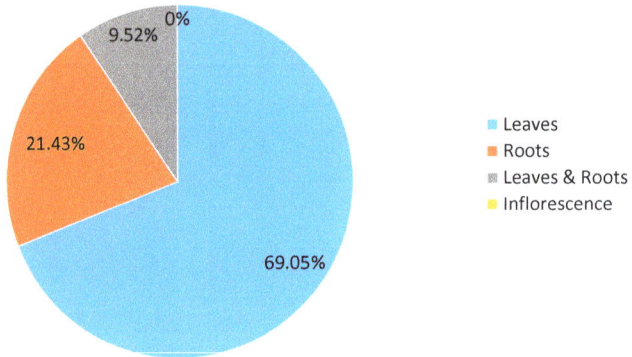

Figure 4. The used part of *Salvia argentea*.

2.1.6. Method of Preparation and Administration

In Saïda, herbalists advocate several ways of preparing *Salvia argentea* for the treatment of respiratory diseases. Powder preparation is the most frequent mode (59.53%), followed by decoction (30.95%) and infusion (9.52%) (Figure 5). All the herbalists (100%) interviewed confirmed that the administration is exclusively oral. The best use of a plant would be that which preserves all its properties while allowing the extraction and assimilation of active compounds [17]. In addition, medicinal plants have side effects when they are incorrectly used by patients [12]. As a result, soft medicine must be practiced with care [18].

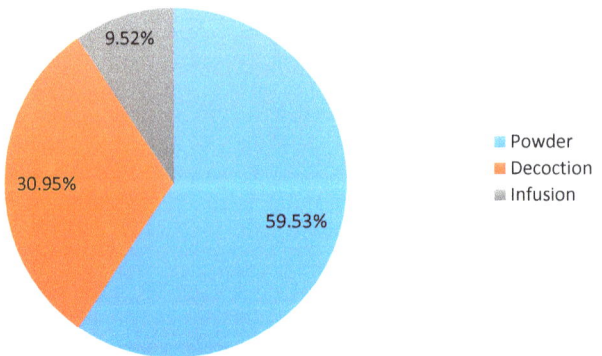

Figure 5. Method of preparation.

2.2. Physicochemical Characterization of the Leaf Powder of Salvia argentea

The results related to the physicochemical analyses of the leaf powder of *Salvia argentea* are reported in Table 1.

Table 1. Physicochemical parameters of the leaf powder of *Salvia argentea*.

Parameters	Powder Leaves of *Salvia argentea*
Humidity (%)	12.89 ± 1.09
Ash (%, Dry Basis)	17.61 ± 0.54
pH	8.05 ± 0.07
Titratable acidity (%)	0.74 ± 0.01

The values obtained for the water content are 12.89% on average. This low content assures that the powder of leaves of *Salvia argentea* can be preserved for a long time without great risk of alterations due to microbial contamination [19].

Total ash is the residue of mineral compounds remaining after incineration of a sample containing organic substances of animal, plant or synthetic origin. Total ash content represents about 17.61% on average of the dry mass. These values are comparable to those found with *Nasturtium officinale* (14.9–17.2%) and *Spinacia oleracea* (18.0–19.1%) [20].

The mean value obtained for the pH is 8.05. This value can be explained by the chlorophyll content of the leaves of *Salvia argentea*, which tends to confer basicity. The values of titratable acidity in term of lactic acid are also correlated with the pH value determined. Their average value is 0.74%. The same trend was observed by Houndji et al. in the leaf powder of *Moringa oleifera* (Lam.) [21].

2.3. Phytochemical Screening

The results of the phytochemical screening are presented in Table 2. They are classified according to various observation criteria, among others: very positive reaction (++++); positive reaction (+++); moderately positive reaction (++); doubtful reaction (+); and negative reaction (−).

Table 2. Phytochemical screening results of the leaf powder of *Salvia argentea*.

Chemical Groups	Results
Alkaloids	++++
Free flavonoids	−
Anthocyanins	++++
Gallic Tannins	++++
Cathechol tannins	++++
Sterols and Terpens	+++
Coumarins	++
Saponins	++++
Oses and holosides	+++
Cyanogenetic derivatives	−

The phytochemical screening carried out on the powder of the leaves of *Salvia argentea* shows the presence of chemical groups which possess interesting biological activities. These include alkaloids, anthocyanin flavonoids, saponins, coumarins, sterols and triterpenes, tannins (gallic and catechic acids) and reducing sugars. The complete absence of cyanogenetic derivatives greatly reduces the toxicological risks associated with the use of *Salvia argentea*.

The presence of potentially active chemical groups such as polyphenolic substances such as tannins in their two forms and anthocyanins in the powder of the studied leaves could justify the traditional indications of this plant by herbalists (survey) in traditional medicine, particularly for their pharmacological properties in the treatment of respiratory diseases [22,23]. Similarly, alkaloids and saponins have been recently reported to be helpful in fighting common pathogenic strains [24]. This plant is therefore a material of choice to enrich the conventional medicine with its interesting biological activities.

3. Materials and Methods

3.1. Biological Material

The *Salvia argentea* samples used in the characterization section were harvested in the Saïda region in 2016, specifically in the Youb region. Plant harvesting was done at the full bloom stage (Figure 6). The identification of the plant was made by Prof. Hasnaoui O., botanist in the Department of Biology of the University of Saïda. A specimen of *Salvia argentea* is deposited in the herbarium of the department of biology of the university.

Figure 6. *Salvia argentea* at the full bloom stage.

3.2. Ethnobotanical Investigation

3.2.1. Study Area

Saïda Province, nicknamed the city of waters because of its numerous springs, is located in the Northwest part of Algeria (34°40′0″ N, 0°19′60″ E). With a population of 350.765, Saïda covers an area of 5536.73 km^2 [25]. It is bordered to the north by the Mascara Province, to the south by the El Bayadh Province, to the west by the Sidi-Bel-Abbes Province and to the east by the Tiaret Province (Figure 7). This position gives it a role of relay between the steppe provinces in the south and the Tell provinces in the north; it allows the extension of the biodiversity of the plant species in this province. It contains 16 communes distributed at the level of 6 districts.

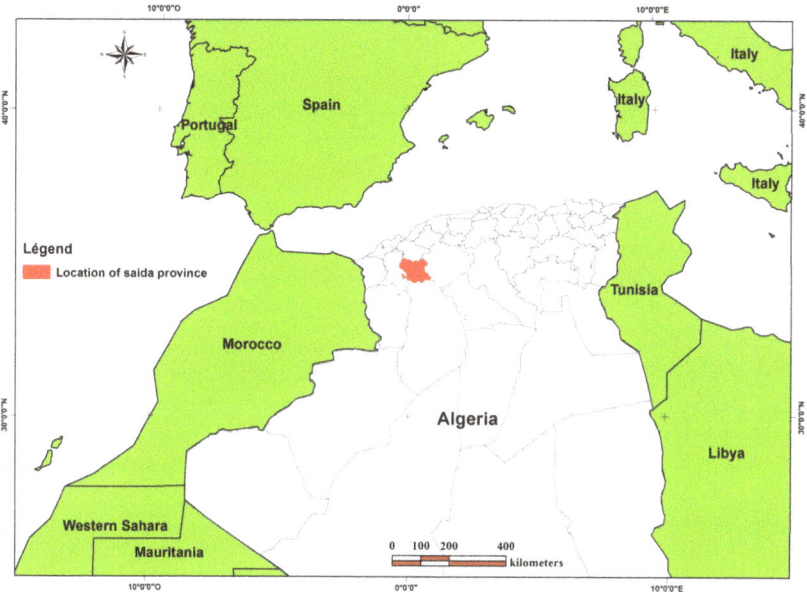

Figure 7. Situation map of Saïda Province.

3.2.2. Methods of Study

The ethnobotanical study was carried out by submitting a questionnaire to 42 herbalists, through 8 communes of the Saïda Province, 14 of them in Saïda, 6 in Youb, 2 in Sidi Boubkeur, 5 in Ain El Hdjar, 2 in Hessasna, 3 in Moulay El Arbi, 4 in Sidi Ahmed and 6 in Ouled Brahim (Figure 8). This disparity reflects the various population densities.

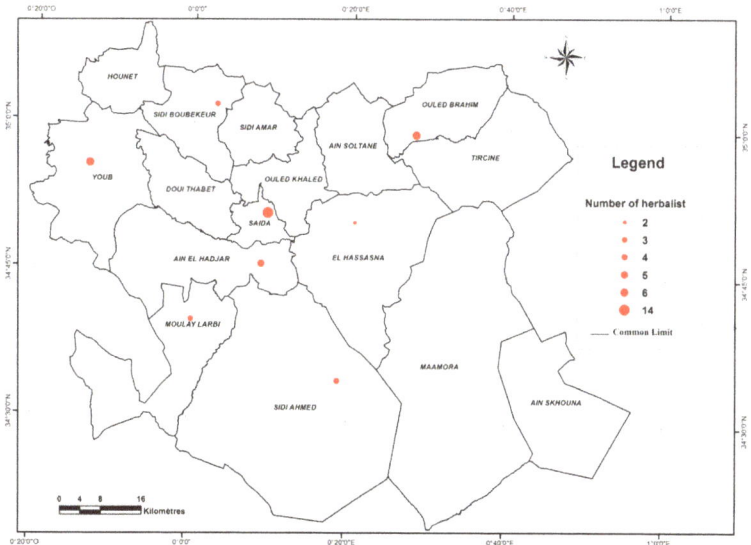

Figure 8. Distribution of survey points in the Saïda Province.

The survey questionnaire form (Appendix A) is divided into 8 parts to evaluate the knowledge of *Salvia argentea* in this area, the use, the prescription and preparation method recommended by each herbalist. All herbalists interviewed were informed about the purpose of this study. The raw data entered on the questionnaire forms were transferred to a database and processed by the Systematic Package for Social Sciences software (SPSS), version 10.

3.3. Characterization of Salvia argentea Leaf Powder

At the end of the ethnobotanical study, a preliminary characterization of some physico-chemical parameters (Humidity, ash, pH and titratable acidity) and phytochemical screening considered as basic analyses were carried out on the powder of the leaves of *Salvia argentea*.

3.3.1. Physico-Chemical Analyses

The humidity content of the previously dried and weighed leaf powder was determined by mass difference before and after desiccation in an oven at +103 °C until a constant mass was obtained [26]. The ash rate was evaluated according to the AFNOR standard NF V 05-104 [27], in which test samples are incinerated at 500 °C until a whitish powder was obtained. The pH and the titratable acidity expressed as a percentage of lactic acid were measured on a suspension made of 10 g of leaf powder in 90 mL of distilled water according to the method described by Nout et al. [28].

3.3.2. Phytochemical Screening

It is a qualitative analysis based on color and/or precipitation reactions which makes it possible to establish the presence or absence of certain bioactive chemical compounds in the plant from its powder. Screening helps to look for: alkaloids, tannins (gallic and catechic), flavonoids (free flavonoids and anthocyanins), reducing compound, coumarins, saponins, sterols and tri-terpenes and cyanogenetic derivatives. These tests are carried out in the presence of certain characterization reagents according to the methods described by Harborne and Bruneton [29,30]:

1. Characterization of alkaloids

The presence of alkaloids is established by salt precipitation and revelation with Mayer's reagent (potassium tetra-iodomercury solution).To 10 g of powder are added 50 mL of 10% H_2SO_4. After 24 h maceration at room temperature, the macerated material is filtered and washed with water to obtain 50 mL of filtrate. To 1mL of filtrate are added 5 drops of Mayer reagent and the mixture is left 15 min at room temperature. The presence of alkaloids is illustrated after a rapid extraction with chlorinated solvent ($CHCl_3$). A sensible quantity of filtrate is made alkaline by dilution with 50% NH_4OH and the same volume of chlorinated solvent is added. After stirring, the organic phase is removed; the remaining is filtered and then evaporated to dryness. Two milliliters of an acid solution (HCl or H_2SO_4) at 10% are added to the residue obtained and the mixture is poured into two test tubes. Five drops of Mayer's reagent are added to the first tube, the second tube serves as a control. The appearance of a white-yellow or light-yellow precipitate confirms the presence of alkaloids.

2. Characterization of Tannins

The presence of gallic and catechic tannins has been demonstrated using ferric chloride. Five grams of sample powder are added to 100 mL of boiling water. After 15 min, the suspension is filtered and rinsed. This infusion will also serve to characterize the presence of flavonoids. Hydrolysable gallic tannins are evidenced by adding 15 mL of Stiasny reagent to 30 mL of the 5% infusion. After heating in a water bath at 90 °C during 15 min, the mixture is filtered and saturated with 5 g of sodium acetate, and then 1 mL of a solution of 1% $FeCl_3$ is added. The appearance of a blue-black tint indicates the presence of gallic tannins. The non-hydrolyzable catechin tannins are characterized by the addition of 1 mL of conc. HCl to 5 mL of the previously prepared infusion. The mixture is boiled for 15 min. In the presence of catechin tannins, a red precipitate, insoluble in isoamyl alcohol, is formed. Catechin tannins are also evidenced after the addition of Stiasny's reagent by the formation of a precipitate.

3. Characterization of flavonoids

The reaction with cyanidine reveals the presence of free flavonoids. To 5 mL of the former 5% infusion, are added 5mL of a mixture of ethanol and concentrated HCl (50:50, v/v %), followed by 1 mL of isoamyl alcohol and a few magnesium chips; the appearance of a pink-orange or purplish pink color reveals the presence of free flavonoids. Anthocyanins are revealed by mixing 5 mL of the infusion with either 5 mL of 10% H_2SO_4 or 5 mL of 50% NH_4OH. If the color of the infusion is accentuated by acidification and then turns blue in alkaline medium, we can conclude to the presence of anthocyanins.

4. Characterization of saponins

A decoction at 1% is prepared by adding 1 g of powder to 100 mL of boiling water; a slight boiling is maintained for 15 min and then the suspension is filtered. Between 1 and 10 mL of filtrate are added successively to 10 test tubes, the volumes are made up to 10 mL with water. The contents of each tube are shaken during 15 s. The height of the resulting foam is measured 15 min after stirring. The growth index is calculated from the tube number (N) in which the foam height is 1 cm. Im = 1000/N.

5. Characterization of reducing compounds

Several reducing compounds can be detected, by preparing a 10% aqueous decoction from 10 g of plant material powder in 100 mL of water for 15 min. After evaporation to dryness of 5 mL of this 10% decoction, 3 drops of concentrated H_2SO_4 are added followed by 4 drops of a saturated solution of thymol in ethanol. The appearance of a red solution reveals the presence of oses and holosides. Cyanogenic glycosides are often found in plants. They are evidenced by carrying out a suspension of 1 g of plant powder in 5 mL of a solution of the same volume of water and toluene. A filter paper strip soaked with Guignard's reagent (2 g of picric acid and 20 g of sodium carbonate in 200 mL of distilled water) is then deposited in the tube. The appearance of a red color indicates the presence of cyanogenic glycosides.

6. Characterization of sterols and triterpenes

The presence of sterols and triterpenes is demonstrated using concentrated H_2SO_4. An extract is first made from maceration for 24 h of 1 g of sample powder in 20 mL of ether. The extract obtained is also used for the characterization of coumarins. Sterols and triterpenes are evidenced by adding 1 mL of $CHCl_3$ to the 10 mL residue of the evaporated macerate. The solution obtained is divided into two test tubes, then 1–2 mL of concentrated H_2SO_4 are added to the bottom of one of the tubes, the second one serves as a control. The formation of a brownish or purple red ring at the interface reveals their presence.

7. Characterization of coumarins

The presence of coumarins is achieved by evaporating to dryness 5mL of an ethereal extract. Two milliliters of hot water are added and then 1 mL of 25% NH_4OH. The mixture is illuminated with UV light at 366 nm. An intense blue fluorescence indicates the presence of coumarins.

4. Conclusions

Salvia argentea has great potential in Algeria for the treatment of respiratory diseases. Many perspectives and expectations emerge from this study, in both the scientific and the public health domains. Thus, the continuity of this study should allow by in vitro and in vivo pharmacological approaches assessing the therapeutic efficacy attributed to *Salvia argentea* and to help clarify the cellular and subcellular mechanisms involved in the anti-inflammatory effects. Further studies would focus on the secondary metabolites and bioactive substances with the help of methods of extraction and fine characterization and should contribute to a better knowledge of the medicinal flora of the traditional Algerian pharmacopoeia.

Acknowledgments: This study is part of a doctoral research. We would like to thank all herbalists in the Saïda Province that kindly participated in our survey and agreed to provide us with information about their careers and profiles. We would also like to thank K. Nabi and A. Baghdadi for their contributions to this study and Michel Guilloton for help in manuscript editing.

Author Contributions: Y.B., K.H. and K.K. conceived and designed the experiments; Y.B. and K.H. performed the experiments; Y.B., K.H. and K.K. analyzed the data; M.S. contributed reagents/materials/analysis tools; and Y.B. wrote the paper.

Conflicts of Interest: The authors declare no conflict of interest.

Appendix A

1- Link: Saida, Youb, SidiBoubkeur, Hessassna, OuledBrahim, Sidi Ahmed, Moulay El Arbi, Ain El-Hadjar.
2- Informant: Age, Sex, Family situation, Academic level, Type of collector, Origin of information.
3- Use of the plant: Therapeutic, Cosmetic, Others.
4- Type of diseases treated: Dermatological, Respiratory, Cardiovascular, Genito-urinary, Osteo-articular, Metabolic, Digestive tract, Neurological, Others.
5- Used part: Leaves, Stem, Flowers, Fruits, Root, Whole plant, Others
6- Preparation: Powder, Infusion, Decoction, Poultice, Maceration, Cooked, Others.
7- Method of administration: Oral, Massage, Rinsing, Painting, Others.
8- Result: Healing, Improvement or Ineffective.

References

1. Hamburger, M.; Hostettman, K. *Bioactivity in Plants: The Link between Phytochemistry and Medicine*; Masson: Paris, France, 1991.
2. Riccobono, L.; Maggio, A.; Rosselli, S.; Ilardi, V.; Senatore, F.; Bruno, M. Chemical composition of volatile and fixed oils from of *Salvia argentea* L. (Lamiaceae) growing wild in Sicily. *Nat. Prod. Res.* **2016**, *30*, 25–34. [CrossRef] [PubMed]
3. Begnis, C. Etude des Propriétés Pharmacologiques et Des Principes Actifs Des Lamiaceae, à L'exception Des Huiles Essentielles. Ph.D. Thesis, University of Montpellier, Montpellier, France, 1995.
4. Rayouf, M.B.T.; Msaada, K.; Hosni, K.; Marzouk, B. Essential Oil Constituents of *Salvia argentea* L. from Tunisia: Phenological Variations. *Med. Aromat. Plant Sci. Biotechnol.* **2013**, *7*, 40–44.
5. Salah, K.B.H.; Mahjoub, M.A.; Ammar, S. Antimicrobial and antioxidant activities of the methanolic extracts of three Salvia species from Tunisia. *Nat. Prod. Res.* **2006**, *20*, 1110–1120. [CrossRef] [PubMed]
6. Fu, Z.; Wang, H.; Hu, X.; Sun, Z.; Han, C. The Pharmacological Properties of *Salvia* Essential Oils. *J. Appl. Pharm. Sci.* **2013**, *3*, 122–127.
7. Farhat, M.B.; Landoulsi, A.; Chaouch-Hamada, R.; Sotomayor, J.A.; Jordán, M.J. Profiling of essential oils and polyphenolics of *Salvia argentea* and evaluation of its by-products antioxidant activity. *Ind. Crops Prod.* **2013**, *47*, 106–112. [CrossRef]
8. Quezel, P.; Santa, S. *Nouvelle Flore d'Algérie et des Régions Désertiques Méridionales*; CNRS: Paris, France, 1963. (In French)
9. Battandier, J.A.; Trabut, M. *Flore de l'Algérie*; ADOLPHE JORDAN: Paris, France, 1888. (In French)
10. Mossi, A.J.; Cansian, R.L.; Paroul, N.; Toniazzo, G.; Oliveira, J.V.; Pierozan, M.K.; Pauletti, G.; Rota, L.; Santos, A.C.A.; Serafini, L.A. Morphological characterization and agronomical parameters of different species of *Salvia* sp. (Lamiaceae). *Braz. J. Biol.* **2011**, *71*, 121–129. [CrossRef] [PubMed]
11. Baran, P.; Özdemir, C.; Aktas, K. Structural investigation of the glandular trichomes of *Salvia argentea*. *Biologia* **2010**, *65*, 33–38. [CrossRef]
12. Rhattas, M.; Douira, A.; Zidane, L. Étude ethnobotanique des plantes médicinales dans le Parc National de Talassemtane (Rif occidental du Maroc). *J. Appl. Biosci.* **2016**, *97*, 9187–9211. (In French) [CrossRef]

13. Doukkali, Z.; Bouidida, H.; Srifi, A.; Taghzouti, K.; Cherrah, Y.; Alaoui, K. Les plantes anxiolytiques au Maroc. Études ethnobotanique et ethno-pharmacologique. *Phytothérapie* **2015**, *13*, 306–313. (In French) [CrossRef]
14. Bigendako-Polygenis, M.J.; Lejoly, J. *La Pharmacopée Traditionnelle au Burundi. Pesticides et Médicaments en Santé Animale*; Press University Namur: Namur, Belgium, 1990; pp. 425–442. (In French)
15. Poffenberger, M.; McGean, B.; Khare, S.; Campbell, J. *Field Method Manual—Community Forest Economy and Use Pattern: Participatoey Rural Appraisal (PRA) Methods in South Gujarat, India*; Society for Promotion of Wastelands Development: New Delhi, India, 1992; Volume 2.
16. Cunningham, A.B. *Peuples, Parc et Plantes: Recommandations Pour Les Zones à Usages Multiples et Les Alternatives de Développement Autour du Parc Naturel de Bwindi Impénétrable, Ouganda*; Documents de travail Peuples et Plantes n° 4; UNESCO: Paris, France, 1996; p. 66.
17. Dextreit, R. *La Cure Végétale, Toutes Les Plantes Pour se Guérir, Vivre en Harmonie*, 3rd ed.; Editions de la Revue Vivre en Harmonie: Paris, France, 1984; p. 184. (In French)
18. Benlamdini, N.; Elhafian, M.; Rochdi, A.; Zidane, L. Étude floristique et ethnobotanique de la flore médicinale du Haute Moulouya, Maroc. *J. Appl. Biosci.* **2014**, *78*, 6771–6787. (In French) [CrossRef]
19. Mafart, P. *Génie Industriel et Alimentaire, Procédés Physiques de Conservation*, 2nd ed.; Lavoisier: Paris, France, 1996. (In French)
20. Analyses Physicochimiques I. Available online: http://dhaouadiramzi.e-monsite.com/medias/files/analysephysico-chimie.pdf (accessed on 18 August 2017). (In French)
21. Houndji, B.V.S.; Ouetchehou, R.; Londji, S.B.M.; Eamouzou, K.S.S.; Yehouenou, B.; Ahohuendo, C.B. Caractérisations microbiologiques et physico-chimiques de la poudre de feuilles de *Moringa oleifera* (Lam.), un légume feuille traditionnel au Bénin. *Int. J. Biol. Chem. Sci.* **2013**, *7*, 75–85. (In French) [CrossRef]
22. Bouchet, N.; Lévesque, J.; Pousset, J.-L. HPLC isolation, identification and quantification of tannins from *Guiera senegalensis*. *Phytochem. Anal.* **2000**, *11*, 52–56. [CrossRef]
23. Andersen, Q.M.; Markham, K.R. *Flavonoids. Chemistry, Biochemistry, and Applications*; CRC Press: Boca Raton, FL, USA, 2006.
24. Kubmarawa, D.; Ajoku, G.A.; Enworem, N.M.; Okorie, D.A. Preliminary phytochemical and antimicrobial screening of 50 medicinal plants from Nigeria. *Afr. J. Biotechnol.* **2007**, *6*, 1690–1696.
25. Rubrique Monographie Wilaya—Aniref.dz. Available online: http://www.aniref.dz/monographies/saida.pdf (accessed on 18 August 2017). (In French)
26. Multon, J.L.; Bizot, H.; Martin, G. Eau (teneur, activité, absorption, propriétés fonctionnelles). Humidités relatives. In *TechniquesD'analyseet de ContrôleDans les Industries Agro-Alimentaires*; Deymie, B., Multon, J.L., Simon, D., Eds.; Lavoisier-Tec et Doc: Paris, France, 1981; Volume IV. (In French)
27. AFNOR. *Produits Dérivés Des Fruits et Légumes Jus de Fruits*; Lavoisier-Tec et Doc: Paris, France, 1986; pp. 36–41. (In French)
28. Nout, M.J.R.; Rombouts, F.M.; Havelear, A. Effect accelerated natural lactic fermentation of infant food ingredients on some pathogenic microorganisms. *Int. J. Food Microbiol.* **1989**, *8*, 351–361. [CrossRef]
29. Harborne, A.J. *Phytochemical Methods a Guide to Modern Techniques of Plant Analysis*, 3rd ed.; Springer: London, UK, 1998.
30. Bruneton, J. *Pharmacognosie, Phytochimie, Plante Médicinales*, 4th ed.; Lavoisier: Paris, France, 2009. (In French)

© 2017 by the authors. Licensee MDPI, Basel, Switzerland. This article is an open access article distributed under the terms and conditions of the Creative Commons Attribution (CC BY) license (http://creativecommons.org/licenses/by/4.0/).

Review

Qualitative and Quantitative Analysis of Polyphenols in *Lamiaceae* Plants—A Review

Katerina Tzima [1,2], Nigel P. Brunton [2] and Dilip K. Rai [1,*]

1. Department of Food BioSciences, Teagasc Food Research Centre Ashtown, D15 KN3K Dublin, Ireland; Aikaterini.Tzima@teagasc.ie
2. UCD Institute of Food and Health, University College Dublin, Belfield, Dublin D04V1W8, Ireland; nigel.brunton@ucd.ie
* Correspondence: dilip.rai@teagasc.ie; Tel.: +353-1-805-9569

Received: 27 February 2018; Accepted: 22 March 2018; Published: 26 March 2018

Abstract: *Lamiaceae* species are promising potential sources of natural antioxidants, owing to their high polyphenol content. In addition, increasing scientific and epidemiological evidence have associated consumption of foods rich in polyphenols with health benefits such as decreased risk of cardiovascular diseases mediated through anti-inflammatory effects. The complex and diverse nature of polyphenols and the huge variation in their levels in commonly consumed herbs make their analysis challenging. Innovative robust analytical tools are constantly developing to meet these challenges. In this review, we present advances in the state of the art for the identification and quantification of polyphenols in *Lamiaceae* species. Novel chromatographic techniques that have been employed in the past decades are discussed, ranging from ultra-high-pressure liquid chromatography to hyphenated spectroscopic methods, whereas performance characteristics such as selectivity and specificity are also summarized.

Keywords: *Lamiaceae*; herbs; polyphenols; liquid chromatography; mass spectrometry

1. Introduction

Concerns over possible adverse health effects of commonly used synthetic antioxidants such as butylated hydroxytoluene (BHT) or butylated hydroxyanisole (BHA) have driven research interests towards finding antioxidants from natural sources, mainly from commonly consumed foods [1,2]. Terrestrial plants constitute one of the most valuable sources of natural antioxidants in addition to other health-promoting phytochemicals [3]. In particular, herbs and spices have shown strong antioxidant activities owing to their high content of polyphenols [2,4].

Considerable attention has been paid to the bioactive compounds in herbs and spices in an effort to reveal their potential contribution to health and the preservation of food quality [5,6]. Several previous studies have suggested that polyphenols from natural sources could be a potential alternative to the use of synthetic antioxidants [3,4]. These antioxidants have many advantages over their synthetic equivalents including consumer acceptance, and the reduced regulatory requirements based on their safety [7]. Natural antioxidants from various botanical sources have been regularly reviewed by focussing on a single species, genus, origin, popularity, applications, bioactivities, selected phytochemical groups of antioxidants, etc. [2]. For instance, *Lamiaceae*, one of the largest herbal families worldwide (236 genera and approximately 6900–7200 species) [8], has been the subject of numerous studies that demonstrated the high radical scavenging capacity (RSC) of its extracts.

Over the last decade, great effort has been devoted to the development of functional food products that can confer positive health-benefits over and above basic nutrition to consumers [9]. Epidemiological findings as well as scientific data have shown that a diet rich in polyphenols, such as flavonoids and hydroxycinnamic acids, has effective health effects [10–13] and could confer protection against the risks of degenerative diseases, e.g., cardiovascular diseases [12]. Therefore, further studies are essential in streamlining the various stages of novel functional food formulations, through improving their health benefits and assuring antioxidant and antimicrobial safety [3,14].

Polyphenols are a group of small organic molecules synthesised by plants as secondary metabolites [15]. These molecules protect the plants from stresses, such as ultra-violet (UV) radiation, infections, cuts, etc. There are many definitions of polyphenols, but the most widely accepted is that "Compounds exclusively derived from the shikimate/phenylpropanoid and/or the polyketide pathway, featuring more than one phenolic unit and deprived of nitrogen-based functions" [15]. Based on this definition, many compounds commonly referred to as polyphenols would not qualify as polyphenols. For example, quinic acid generally listed with polyphenols, is biosynthesized independent of the shikimate pathway Therefore, it cannot be considered as phenolic acid [16]. In the present review compounds such as those presented in Figure 1 will be referred to as polyphenols. Flavonoids, a subset of polyphenols, are characterized by at least two phenol subunits (Figure 1b). The reactive nature of the polyphenols often leads to conjugation with glucose, cellulose, proteins, and with same or other polyphenols forming oligomers (Figure 1c). Several thousand polyphenols have been reported in higher plants [15] and this structural diversity is one of the factors contributing to the complexity of their analysis [17]. Compounded to this is the huge variation in the levels of these compounds in different plant species [3]. The need for sensitive and accurate methods for the analysis of polyphenols is essential, as knowledge of dosage are prerequisites in evaluating health claims of food components.

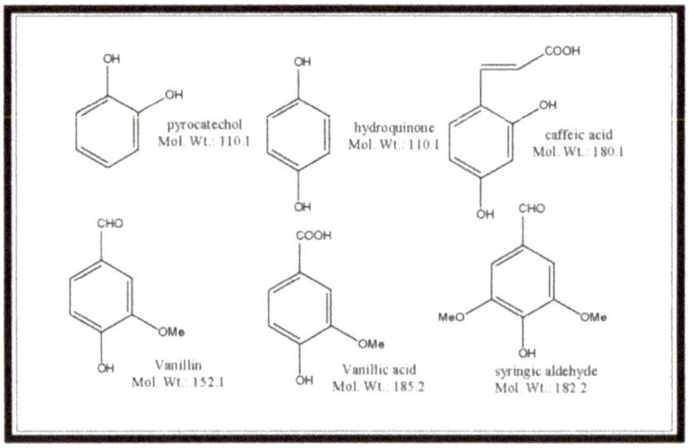

(a)

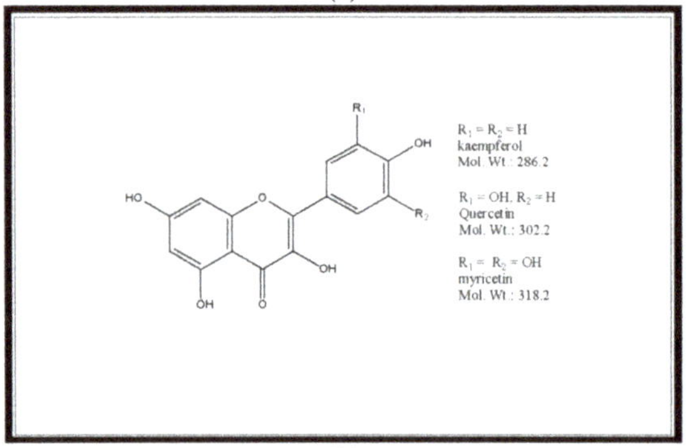

(b)

Figure 1. *Cont.*

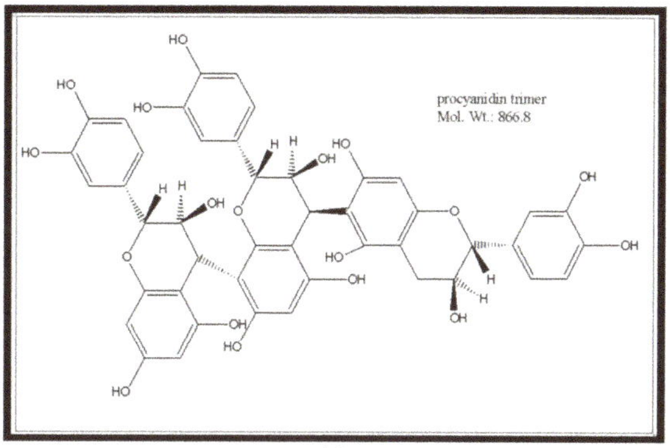

(structure of procyanidin trimer, Mol. Wt.: 866.8)

(c)

Figure 1. Examples of (**a**) simple and (**b**,**c**) complex polyphenols in plants.

Classical techniques such as high-performance liquid chromatography (HPLC), thin layer chromatography (TLC), gas chromatography (GC), and capillary electrophoresis (CE), which rely on UV spectrophotometry as the detection tool, have been used for the analysis of polyphenol profiles in herbs [18]. These methods generally lack specificity and sensitivity and rely on the chemical nature of the analytes (chromophore). A common issue being the interference by plant/biological matrices in the UV-dependent assays such as TLC, CE, and HPLC. This has led to an interest in mass spectrometry (MS) coupled with either liquid chromatography (LC) or GC, which has the added advantages of specificity and sensitivity [19]. This review describes the recent (2013–2018) developments and applications of analytical methods in qualitative and quantitative studies of polyphenols following extraction, with special focus on the *Lamiaceae* spices.

2. Extraction and Purification

The choice and collection of plant tissues constitute the initial steps for the identification and quantification of bioactive compounds [20]. In order for an analytical technique to generate sufficient data for the determination of natural substances such as polyphenols in plants, it must be sufficiently efficient, selective and sensitive [21]. In this regard, sample preparation is a crucial step before analysis [22], while the sensitivity of the analytical technique is dependent on the polyphenol extraction choice, the purification steps, and the initial concentration of polyphenols in the plant crude extracts prior to analysis [23]. Ideally extraction should result in the selective separation of the target components with high recovery and reduced interferences [24]. Extracts can be obtained with several solvents [20], either organic or inorganic, which can determine the quantity of the extracted phenolics [25]. The most crucial aspect that should be considered for the solvent choice is the polarity of the targeted compounds [26]. Nonetheless, various other factors such as extraction time, temperature, extraction steps, solvent-to-sample ratio [25], molecular affinity among solute and solvent, and use of co-solvents [26] may additionally influence the extractability of phenolics [25]. The optimal content of phenolics is also dependent on the nature of the plant matrix and its bioactive constituents [25]. Plant bioactives can be recovered with several conventional extraction methods, including maceration, distillation, Soxhlet extraction [26], reflux extraction [27], and low pressure solvent extraction (LPSE) [28]. However, these techniques are labor-intensive as they require extended extraction times, large quantities of solvents, and they commonly result in low extraction yields and reduced selectivity [29–32]. In parallel, the extracts may be subjected to excessive oxygen (O_2), heat and light, leading to their subsequent degradation [27,29]. Regardless of their inherent multiple drawbacks, liquid-liquid and solid-liquid extraction procedures are still regularly employed [33].

Several novel extraction methods have been established for the recovery of phenolics from plant materials, including microwave-assisted extraction (MAE), supercritical fluid extraction (SFE) [32,34], ultrasound-assisted extraction (UAE) [34] and accelerated solvent extraction (ASE) [29]. In recent years, the use of MAE has gained considerable popularity due to its benefits of diminution of extraction time, reduced cost, sustainability, as well as potential for automation or on-line connection to analytical instrumentation [34–36]. Nonetheless, there are certain drawbacks regarding its use in the recovery of polyphenols, in particular the various parameters that could potentially affect its effectiveness, such as the microwave utilization time and power, surface area of the sample, temperature, nature of sample matrix and sample purity [37]. UAE constitutes one of the most simple and convenient extraction processes employing mechanic vibrations generated by sound waves (>20 kHz) for extracting bioactive compounds [25,32]. Nevertheless, in some cases it has been reported that a prolonged sonication (>40 min) in frequencies above 20 kHz could have a detrimental effect on the targeted components. This effect was ascribed to the reduction of diffusion area and rate, but also the increased diffusion distance, which may lead to minimized yield of total phenolics and flavonoids. Furthermore, a potential formation of free radicals may occur [38]. For ASE extraction techniques, low-boiling solvent/solvent mixtures in parallel to increased temperature (>200 °C) and pressure (3000 psi/206.8 bar) are employed. This results reduced solvent viscosity and tension with a parallel elevation of the solvent diffusion rate, mass transfer, and solubility of the targeted components are accomplished. Compared to conventional extraction techniques, ASE utilizes reduced solvent quantities, is time-efficient and automated, and protects the samples from exposure in O_2 and light [29]. The different characteristics of the SFE extraction process, including the utilization of low temperatures, the absence of O_2, and the common use of carbon dioxide (CO_2) render it as a superior procedure for extracting bioactive components [39]. As CO_2 is economic, non-toxic, nonflammable, and volatile, it may be used in various conditions [40]. In the case of volatile compounds in plant materials such as phenolic terpenes, an extraction process that can be employed is purge and trap (P & T) [41]. This dynamic technique is dependent on bubbling through the sample by using an inert gas such as helium or nitrogen (N_2). Subsequently, the volatile components of the sample are adsorbed on a trap that is directly heated to desorb them into a gas chromatograph injector [42]. The P & T technique is efficient and results in increased extractability [41].

Matrix effects (ME) constitute a significant disadvantage of LC-MS analysis that the matrix can cause suppression or enhancement of ionization, and subsequent quantification errors [43]. Purifications steps are used to eliminate matrix interferences such as lipids, carbohydrates, or undesirable molecules, and involve the removal of interfering components from the crude extract with an adsorption-desorption process or partitionable solvents (chloroform, hexane, dichloromethane) and open column chromatography [19,44]. Amberlite resin and solid phase extraction (SPE) cartridges are also frequently used materials for the purification of phenolics from crude extracts [44]. With the use of SPE, several disadvantages related to liquid-liquid extraction including use of excessive quantities of solvents, incomplete phase separations, and poor recoveries can be overcome [25]. Although, SPE is commonly employed for the removal of non-phenolic compounds such as sugars, organic acids, and other water-soluble components, this will also lead to the loss of highly polar phenolics [11,44,45]. In addition, there are also costs involved on the SPE manifold and the associated consumables [25]. Table 1 summarizes the extraction processes that were found in the recent literature, regarding the extraction of (poly) phenolic compounds from *Lamiaceae* herbs prior to chromatographic analysis.

Table 1. Extraction processes for polyphenolic constituents from *Lamiaceae* herbs.

i) *Lamiaceae* Species ii) Plant Part	Extraction Process	Polyphenol Classes	i) Solvent ii) Solute: Solvent Ratio	i) Time (t) ii) Temperature (T)	i) Work-up and Conditions ii) Purification/Clean-up [1]	Reference
i) *Mentha pulegium*; *Nepeta nuda* ii) Aerial parts	Reflux condensation	Phenolic acids; Flavonoids	i) methanol (MeOH) ii) 1:10 weight/volume (w/v)	i) 30 min ii) not available (n/a)	i) Exhaustive-extraction (two times); Filtration ii) n/a	[46]
i) *Thymus vulgaris* ii) Aerial parts	Reflux (hot) extraction	Flavonoids (flavones)	i) MeOH ii) 1:6 (w/v)	i) n/a ii) n/a	i) 3 Extraction Repetitions; Drying (rotary evaporator); Reconstitution of residue (1.5 g residue: 5 mL MeOH); Filtration; Dilution (1:2) with 0.5 mL borax buffer (20 mM, pH 10.0) ii) n/a	[47]
i) 11 species of *Mentha*; 2 Mixtures of *Mentha* species ii) Plant material; Pharmaceutical products	Soxhlet extraction of residue after chlorophyll removal	Hydroxycinnamic acids; Flavonoids	i) MeOH ii) 1:10 (w/v)	i) 8 h ii) n/a	i) Evaporation (water bath, 0.9 atm); Dissolution of residue to 25 mL with MeOH ii) Isolation of Chlorophylls: Soxhlet extraction with chloroform, 8 h, 20 g of solute	[48]
i) *Melissa officinalis* ii) Fresh herbs or leaves	Sonication	Hydroxybenzoic, Hydroxycinnamic acids	i) 80% aqueous MeOH ii) 1:8 (w/v)	i) 30 min ii) ambient	i) Centrifugation (20,000 rpm, 10 min); Two process repetitions; Combination of extracts; Dilution (Final volume: 25 mL, with 80% aqueous MeOH); Filtration ii) n/a	[49]
i) *Origanum vulgare* ssp. *hirtum*; *Thymus capitatus*; *Satureja thymbra*; *Melissa officinalis*; *Rosmarinus officinalis* ii) Aerial parts, dried, grounded leaves and flowers	Sonication	Phenolic acids and their derivatives; Flavonoids; Phenolic monoterpenes	i) 70% aqueous MeOH or water (H$_2$O) ii) 1:8 (w/v)	i) 20 min ii) ≤30 °C	i) Centrifugation (12,500 rpm, 15 min, 4 °C); Filtration ii) n/a	[50]
i) *Rosmarinus officinalis*; *Origanum vulgare*; *Thymus vulgaris*; *Origanum majorana* ii) Dried, grounded	Sonication	Flavonoids; Phenolic acids; Phenolic terpenes	i) 0.1% formic acid in 50% aqueous ethanol (EtOH) ii) 1:5 (w/v)	i) 5 min ii) n/a	i) Centrifugation (3000 g, 10 min, 4 °C); Two repetitions (residue); Combination of extracts; Evaporation with N$_2$; Reconstitution of extracts to 5 mL with 0.1% aqueous formic acid ii) Solid-Phase Extraction (SPE: Dilution (1 mL extract, 1 mL H$_2$O, 34 µL 35% hydrochloric acid (HCl)); Equilibration (1 mL MeOH, 1 mL sodium acetate 50 mmol/L, pH 7); Rinsing (sodium acetate 50 mmol/L, pH 7.5% MeOH); Elution of polyphenols (1800 µL 2% formic acid in MeOH); Evaporation (N$_2$); Residue dilution to 250 µL with 1% formic acid in H$_2$O); Filtration	[1,51]

Table 1. Cont.

i) *Lamiaceae* Species ii) Plant Part	Extraction Process	Polyphenol Classes	i) Solvent ii) Solute: Solvent Ratio	i) Time (t) ii) Temperature (T)	i) Work-up and Conditions ii) Purification/Clean-up [1]	Reference
i) *Mentha pulegium*; *Origanum majorana* ii) Aerial parts	Sonication	Flavonoids; Hydroxybenzoic, Hydroxycinnamic acids and their derivatives	i) MeOH ii) 1:10 (w/v)	i) 30 min ii) ambient	i) Centrifugation (3500 rpm, 10 min); four repetitions; Collection of supernatants; Evaporation (reduced pressure, 35 °C); Residue re-constitution to 2 mL with MeOH; Filtration ii) n/a	[52]
i) *Rosmarinus officinalis* ii) Branded extract rich in carnosic acid	Sonication	Flavonoids (mainly flavones); Phenolic terpenes (diterpenoids and derivatives); Phenolic acids	i) 2% formic acid in acetonitrile (MeCN) ii) 1:6.7 volume/volume (v/v)	i) 10 min ii) n/a	i) Centrifugation (10,480 g, 5 min, Ambient T); Direct injection after centrifugation ii) n/a	[53]
i) 6 *Ocimum* spp. ii) Leaves, dried, grounded	Sonication (53 kHz)	Phenolic acids; Flavonoids; Propenyl phenols; Terpenoids	i) 80% aqueous MeOH ii) 1:10 (w/v)	i) 30 min ii) ambient	i) Maintenance 24 h (22–24 °C); Filtration; Evaporation (reduced pressure, 40 °C) ii) Sonication of the residue (1 mg) in MeCN (1 mL) Filtration (0.22 µm filter); Dilution to 30 ng/mL (MeCN); Spiking (andrographolide).	[54]
i) *Satureja montana* ssp. *kitaibelii* ii) Aerial parts of wild plant, air-dried, milled	Solid-liquid extraction, Sonication	Hydroxybenzoic, Hydroxycinnamic acids; Phenyl acetic acids; Flavonoids (flavones, flavonols)	i) 60%, 70%, and 80% aqueous MeOH, EtOH and acetone ii) 1:10 (w/v)	i) 10 min ii) n/a	i) Centrifugation (1000 g, 15 min); Removal of supernatant and exhaustive extractions (three repetitions); Evaporation of supernatants; Reconstitution in MeOH: H_2O 50:50 (v/v) (1 mL); Filtration ii) n/a	[55]
i) *Mentha spicata* ii) Commercial extract	Solid-liquid extraction, Sonication	Hydroxybenzoic, hydroxycinnamic acids; Flavonoids (flavones, flavonols)	i) 80% aqueous MeOH with 1% formic acid ii) 1:5 (w/v)	i) 25 min ii) ambient	i) Centrifugation (10,480 g, 5 min, ambient T); Exhaustive extraction (three repetitions: on the same sample) ii) n/a	[56]
i) 3 *Mentha* spp. ii) Dried and powdered leaves	Solid-liquid extraction of defatted residues	Phenolic acids; Flavonoids	i) EtOH ii) 1:40 (w/v)	i) 24 h ii) ambient	i) Filtration (on cellulose); Concentration (vacuum evaporator, 40 °C) ii) Defatting; Stirring (130 rpm); 25 g of sample in *n*-hexane (600 mL); Ambient T; 3 h	[57]
i) *Origanum vulgare*; *Ocimum basilicum*; *Rosmarinus officinalis*; *Origanum majorana*; *Thymus vulgaris*; *Satureja hortensis* ii) Commercial, dried, grounded leaves	Shaking, Solid-liquid extraction	Phenolic acids	i) 70% aqueous EtOH ii) 1:10 (w/v)	i) 2 h ii) ambient	i) Filtration; Vacuum evaporation (40 °C); Freeze-drying; Analysis concentration: 0.1% (w/v) ii) n/a	[2]

Table 1. Cont.

i) *Lamiaceae* Species ii) Plant Part	Extraction Process	Polyphenol Classes	i) Solvent iii) Solute: Solvent Ratio	i) Time (t) ii) Temperature (T)	i) Work-up and Conditions ii) Purification/Clean-up [1]	Reference
i) *Thymus vulgaris*; *Salvia officinalis* ii) Aerial parts	Maceration (herbal tinctures)	Phenolic acids (hydroxycinnamic acids); Flavonoids (flavonols, flavones)	i) 70% aqueous EtOH ii) n/a	i) 7 days ii) n/a	i) (According to the Polish Pharmacopoeia VI protocol) ii) n/a	[58]
i) *Thymus x citriodorus* ii) Mixture of leaves and stems, dried	Maceration of residue (defatted)	Phenolic acid derivatives; Flavonoids (flavones, flavonols, flavanones)	i) 80% aqueous EtOH ii) 1:60 (w/v)	i) 30 min ii) ambient	i) Filtration; Four re-extractions of residue; Combination of extracts; Lyophilization ii) Defatting: Maceration with *n*-hexane (150 mL); 5 g of sample; 30 min; Ambient (T); three repetitions	[59]
i) *Origanum majorana* ii) Commercially produced, dried, grounded	Solid-liquid extraction	Flavonoids; Phenolic acids	i) 80% MeOH ii) 1:10 (w/v)	i) 6 h, 16 h ii) 23 °C	i) Filtration; Combination of extracts; Drying (rotary evaporator, 50 °C); Dissolution in H_2O (16.5 g/500 mL) ii) Liquid-liquid partitioning for flash chromatography (FC): ethyl acetate (AcOEt) (500 mL) in H_2O (500 mL) with 16.5 g of extract; Dissolution of polar part (14.7 g) in H_2O (50 mL) and non-polar part (1.7 g) in AcOEt (50 mL)	[60]
i) *Rosmarinus officinalis*, *Origanum majorana*, *Origanum vulgare Ocimum basilicum*, *Mentha spicata*, *Thymus vulgaris Mentha x piperita*, *Thymus x citriodorus* ii) Fresh; Dried; Organic dried	Solid-liquid extraction aided by shaking	Hydroxybenzoic, hydroxycinnamic acids; Flavonoids; Phenolic terpenes	i) MeOH ii) 1:100 (w/v)/1:12.5 (fresh) (w/v)	i) 10 min ii) n/a	i) Centrifugation (2000 rpm, 10 min); Residue re-extraction (initial conditions); Combination of supernatants; Evaporation (40 °C, Final Volume: 5 mL); Dilution to 10 mL with MeOH ii) n/a	[14]
i) *Origanum vulgare* ii) Herb sample from 2 different sources, dried	Solid-liquid extraction aided by shaking (Soluble, Bound extracts)	Hydroxycinnamic, hydroxybenzoic acids;Phenolic monoterpenes (Soluble extracts) Hydroxycinnamic, hydroxybenzoic acids (Bound extracts)	i) 80% aqueous MeOH (Soluble extracts); 2 M sodium hydroxide (NaOH) (Bound extracts) ii) 1:20 (w/v) (Soluble extracts); n/a (Bound extracts)	i) 24 h (Soluble extracts); 4 h (Bound extracts) ii) ambient	i) Soluble extracts: Centrifugation (2000 g, 30 min, Ambient T); Supernatant and soluble fraction collection Bound extracts: pH 2.0 with 6 M HCl; Centrifugation (2000 g, 30 min, ambient T); Collection of supernatant; Extraction (15 mL 1:1 (v/v) Diethylether: AcOEt-three times); Evaporation of organic layers (30 °C); Dissolution to 10 mL with 80% aqueous MeOH ii) n/a	[13]
i) Sicilian *Origanum vulgare* ssp. *hirtum*, *Rosmarinus officinalis*, *Thymus capitatus* L. ii) Dried-aerial parts, flowering season samples from various sites	Solid-liquid extraction (Nonvolatile fraction); Hydrodistillation (Volatile fraction)	Flavonoids (flavones, flavanones) (Nonvolatile fraction); Phenolic terpenes (Volatile fraction)	i) AcOEt and EtOH (Nonvolatile fraction); n/a (Volatile fraction) ii) 1:6.7 (w/v) (3 times) (Nonvolatile fraction); n/a (Volatile fraction)	i) Overnight in the dark (Nonvolatile fraction); 3 h (Volatile fraction) ii) ambient	i) Nonvolatile fraction: Storage: 4 °C, N_2-rich atmosphere; Analysis concentration: Dissolution of 10–20 mg of each sample in MeOH (1.5 mL); Filtration. Volatile fraction: (According to European Pharmacopoeia); Drying with sodium sulfate anhydrous (Na_2SO_4); Storage: under N_2 ii) Nonvolatile fraction: Defatting with *n*-hexane; 30 g dried, grounded aerial parts/200 mL; 3 times	[61–63]

Table 1. Cont.

i) *Lamiaceae* Species ii) Plant Part	Extraction Process	Polyphenol Classes	i) Solvent ii) Solute: Solvent Ratio	i) Time (t) ii) Temperature (T)	i) Work-up and Conditions ii) Purification/Clean-up [1]	Reference
i) *Thymus serpyllum* ii) Whole-dried	Solid-liquid extraction (Phenolic fraction); Purge & Trap (N$_2$, 500 mL N$_2$/min) followed by SPE (Volatile fraction)	Flavonoids; Phenolic acids; Phenolic terpenes (monoterpenes)	i) 75% aqueous MeOH (Phenolic fraction); adsorbent: Lichrolut EN (Volatile fraction) ii) 1:4 (w/v) (Phenolic fraction); 3 g/200 mg (Volatile fraction)	i) 2hr (Phenolic fraction); 90 min (Volatile fraction) ii) n/a	i) Phenolic fraction: Residue washing (5 mL of 75% aqueous MeOH); Combination of extracts; Filtration; Vacuum evaporation (20 °C). Volatile fraction: Elution (Dichloromethane); Dehydration (Anhydrous Sodium Sulphate); Concentration (5 mL, Snyder column, 40 °C); Re-concentration to 0.5 mL (N$_2$); Filtration ii) n/a	[64]
i) *Mentha australis* R. Br ii) Fresh leaves and stems	Solid-liquid extraction following sonication	Phenolic acids; Flavonoids (flavanone glycosides)	i) 80% aqueous MeOH ii) 1:20 (w/v)	i) 10 min, 2 h; overnight ii) 4 °C	i) Extraction 1: Centrifugation (10,000 g, 15 min). Extraction 1, 2, 3: Combination of supernatants; Solvent evaporation (vacuum rotary evaporator, 40 °C) ii) Purification: Glass column (25 × 300 mm i.d.); 50 mL extract; Addition of Amberlite resin; Washing with H$_2$O; Elution with 80% aqueous MeOH; Vacuum evaporation (40 °C); Lyophilization (−109 °C, 0.015 k Pa); Analysis concentration: 1 mg (lyophilized, purified) extract/mL MeOH	[65]
i) 3 species of *Salvia* ii) Aerial parts, dried, pulverized	Solid-liquid extraction of the residue obtained after removal of lipophilic substances	Flavonoids (flavones, flavone glycosides)	i) Hot H$_2$O (~90 °C) ii) 1:40 (w/v)	i) Left to reach ambient (T) ii) n/a	i) Partitioning (3 × 100 mL AcOEt, 3 × 100 mL *n*-butanol); Combination of organic phases; Drying (anhydrous magnesium sulfate); Drying (rotary evaporator, 40 °C; Dissolution to 3 mL with MeOH ii) Lipophilic content removal: Shaking (5 g of pulverized sample in *n*-hexane (100 mL), 30 °C, 2 h); Filtration; Stirring overnight (30 °C, 100 mL MeOH: dichloromethane 1:1); Filtration; Drying (rotary evaporator, 40 °C)	[66]
i) *Rosmarinus officinalis* ii) Leaves from 20 geographical zones	Microwave assisted extraction (MAE); two pre-heating steps (160 and 320 W); two extraction cycles (800 W)	Flavonoids; Phenolic diterpenes	i) 70% aqueous MeOH ii) 1:12.5 (w/v)	i) Each pre-heating step:1 min; Heating gaps: 15 s; Each extraction cycle: 5 min ii) n/a	i) Combination of extracts (two extraction cycles); Filtration; Evaporation (rotary evaporator); Analysis concentration: 800 μg/mL in 50% aqueous MeOH; Filtration ii) n/a	[67]
i) (a) *Origanum majorana*; (b) *Mentha pulegium*; (c) *Lavandula officinalis* ii) (a) Leaves and aerial parts; (b) Flowers; (c) Leaves, dried, milled	MAE (500 W)	Flavonoids; Hydroxycinnamic, hydroxybenzoic acids	i) 60 and 80% aqueous MeOH, EtOH and acetone ii) 1:15 (w/v)	i) 15 min ii) 80 °C	i) Irradiation process: 3 min heating for reaching 80 °C, 3 min for balancing at 80 °C, 5 min for cooling; Filtration ii) n/a	[68]

Table 1. Cont.

i) *Lamiaceae* Species ii) Plant Part	Extraction Process	Polyphenol Classes	i) Solvent ii) Solute: Solvent Ratio	i) Time (t) ii) Temperature (T)	i) Work-up and Conditions ii) Purification/Clean-up [1]	Reference
i) *Rosmarinus officinalis*; *Salvia officinalis*; *Origanum vulgare*; *Thymus vulgaris* ii) Leaves, or herbalmix, or as ingredients in chimichurri sauce	Supercritical fluid extraction—carbondioxide (SFE-CO_2); Soxhlet Low Pressure Solvent Extraction (LPSE) (17.3 g/min); Ultrasound assisted extraction (UAE) (40 kHz;1 bar; 20 g of CO_2/g raw material solvent)	Phenolic terpenes (diterpenes)	i) CO_2 for SFE; EtOH for Soxhlet LPSE and UAE; ii) n/a for SFE and UAE; 1:30 for Soxhlet and UAE.	i) 6 h ii) 40 °C for SFE and S; n/a for Soxhlet; 50 °C for UAE	i) n/a for SFE; Vacuum evaporation (40 °C) for Soxhlet and UAE ii) n/a	[69]
i) 10 *Salvia* species ii) Plant material, dried	SFE-CO_2 (45 MPa, CO_2; 2 L/min) Accelerated solvent extraction (ASE) (10.3 MPa)	Flavonoids; Phenolic terpenes; Hydroxybenzoic, hydroxycinnamic acids; Phenolic acids (caffeic acid derivatives)	i) CO_2 (99.9%) for SFE; 96% EtOH, followed by H_2O for ASE ii) n/a for SFE; 3:1 in diatomaceous earth for ASE	i) 60 min. (SFE-CO_2); 30 min. (ASE) ii) 60 °C for SFE; 140 °C for ASE	i) ASE: EtOH evaporation; Lyophilization of H_2O extracts ii) n/a	[3]
i) *Salvia officinalis*, *Thymus serpyllum*, *Origanum vulgare*, *Melissa officinalis* ii) Plant raw material, grounded	Heating; MAE; Sonication; Subcritical extraction	Phenol carboxylic; Cinnamic acids; Flavonoids; Phenolic terpenes (diterpenes)	i) 70% aqueous EtOH ii) 1:50 (*w/v*)	i) n/a ii) n/a	i) (According to the Russian State Pharmacopoeia, FS.2.5.0051.15). Centrifugation; Filtration ii) n/a	[70]

[1] Purification/Clean-up step took place either in parallel or subsequently to the extraction of (poly) phenolic/bioactive compounds.

3. Chromatographic Techniques with Ultraviolet/Visible (UV/Vis) Based Detection

Chromatography, in particular HPLC, is still the most widely used analytical tool for the identification and quantification of polyphenols, which are inherently chromophoric in nature [17,71,72]. In LC, some characteristics of eluted polyphenols can be archived using the detection system, depending on the chemical structure of the molecule. For example, UV/Vis absorption spectra in parallel to the retention time can, with the use of authenticated standards, contribute to the identification of polyphenols in *Lamiaceae* herbs [72].

The separation of phenolics has been improved with the use of reversed-phase (RP) columns (mainly RP C18); however C8 and C12 columns have also been investigated in herbal analysis [73–75]. Typical C18 columns in most of the reported HPLC analysis are 100–200 mm length, internal diameters of 3.9–4.6 mm, and stationary phase particle sizes equal to 3–10 μm [23]. A summary of recently reported researches employing conventional as well as hyphenated chromatographic techniques for the qualitative and quantitative analysis of (poly) phenolic compounds in *Lamiaceae* herbs is presented in Table 2.

Regarding the eluents, organic solvents such as MeOH or MeCN in conjunction with aqueous solvents are used [19]. The use of a H_2O/MeCN binary rather than H_2O and MeOH did not show any significant improvement in resolution on the HPLC separation of phenolic acids of methanolic extracts of lemon balm (*Melissa officinalis*) (Table 2). Thus, a combination of H_2O and MeOH could be used to eliminate the cost and toxicity restrictions of MeCN [49]. Elimination of peak tailing in phenolic profile analysis is achieved through the use of various buffers [19] for eluent acidification, as for instance TFA [49], acetic, formic or phosphoric acids, with concentrations ranging from 0.01% to 6% to be the most frequently reported [19]. In addition to the choice of columns and solvents, a significant parameter that influences the separation of phenolic compounds in chromatography is the column temperature [73]. High temperatures lead to reduced eluent viscosity, resulting in shorter elution times, and thus decreasing the organic solvent consumption [17]. As it has been revealed, a temperature of 30 °C gave rise to improved chromatographic resolution of phenolic acids in *Melissa officinalis* (Table 2), compared to 20 °C and 25 °C [49]. Nonetheless, the maximum column functional temperature is 60 °C, whereas higher temperatures could significantly decrease the estimated column life time [69] and may lead to thermal degradation of targeted polyphenols. Therefore, a column temperature equal to 55 °C was used in the research of Zabot et al. [69] to identify phenolic terpenes in different herbs (Table 2). This study had shown that elevating temperature led to a proportional mean reduction of the retention times of the analytes, and accordingly to lower peak widths, increased peak height and an enhanced chromatographic resolution [69].

Table 2. Recent applications of conventional and hyphenated chromatographic methods for phenolic constituents in *Lamiaceae* species.

i) *Lamiaceae* Species ii) Plant Part	Polyphenols Analysed [1]	Chromatography	Detection System	Chromatographic Conditions and Method Validation Results	Reference(s)
i) *Thymus vulgaris* ii) Aerial parts	C17, **C21**	Capillary electrophoresis (CE)	UV-diode array detector (DAD)	Capillary: Fused silica (66 cm length, 58 cm effective length, 75 mm internal diameter (i.d.)) Capillary (T): 23 °C Background electrolyte solution: borax buffer (20 mM, pH 10.0): 90% MeOH Driving voltage: 23 kV limit of detection (LOD) for C17: 0.53 µg/mL, LOD for C21: 1.05 µg/mL limit of quantification (LOQ) for C17: 1.41 µg/mL, LOQ for C21: 2.98 µg/mL correlation/determination coefficient (R^2) for C17: 0.9990, (R^2) for C21: 0.9999	[47]
i) *Melissa officinalis* ii) Fresh herbs or leaves from 12 manufacturers	C57, C59, **C63**, C64, C66, C67	High performance liquid chromatography (HPLC)	UV/Vis	Column: Hypersil GOLD C18 (250 mm × 4.6 mm i.d., 5.0 µm particle size (p.s.)) (T): 30 °C Eluents: (A) 0.05% trifluoroacetic acid (TFA) in MeOH; (B) 0.05% TFA in H_2O Run (t): 35 min LOD: 0.16–0.51 µg/mL, LOQ: 0.42–1.54 µg/mL, (R^2): ≥0.9089	[49]
i) *Mentha pulegium, Nepeta nuda* ii) Aerial parts	**C17**, C19, C21, C22, C33, C41, C44, C45, C46, **C57**, C59, C64	HPLC	UV-photodiode array (PDA) detector	Column: LiChrospher 100 RP C18 endcapped (250 mm × 4.6 mm i.d., 5.0 µm p.s.) Eluents: (A) H_2O containing 0.02% phosphoric acid and (B) MeCN Run (t): 70 min	[46]
i) *Origanum vulgare* ii) Herb sample from different sources, dried	C15, C16, C34, C36, C38, C55, C56, **C57**, C58, C59, C61, C63, **C66**, C69, C71, **C75**	HPLC	DAD	Column: Zorbax SB-Aq (250 mm × 4.6 mm i.d., 5.0 µm p.s.) Eluents: (A) 0.5% formic acid in H_2O; (B) MeOH Run (t): 95 min	[13]
i) *Salvia officinalis, Thymus serpyllum, Origanum vulgare, Melissa officinalis* ii) Plant (raw) material	C22, **C23**, C46, C57, C58, C59, C61, **C63**, C64, C65, C66, C68, C69, C70, C78	HPLC	DAD	Column: Phenomenex Luna C18 (250 mm × 4.6 mm i.d., 5.0 µm p.s.) (T): 40 °C Eluents: (A) MeCN; (B) 1% acetic acid in H_2O Run (t): 35 min LOD: 0.10–0.30 µg/mL, (R^2): ≥ 0.999	[70]
i) *Origanum vulgare* ssp. *hirtum, Thymus capitatus, Satureja thymbra, Melissa officinalis, Rosmarinus officinalis* ii)Aerial parts, dried, grounded leaves and flowers	C1, C17, C21, C34, C36, **C37**, C40, C46, C48, C50, C57, C58, C59, C60, C61, **C63**, C64, C66, C67, C68, C69, C70, C74, **C75**	RP-HPLC	DAD	Column: Nucleosil 100 C18 (250 mm × 4.6 mm i.d., 5.0 µm p.s.) (T): 30 °C Eluents: (A) 1% acetic acid in H_2O; (B) MeCN; (C) MeOH Run (t): 55 min LOD: 0.002–0.16 µg/mL, LOQ: 0.01–0.48 µg/mL, (R^2): ≥ 0.9961	[50]

Table 2. Cont.

i) Lamiaceae Species ii) Plant Part	Polyphenols Analysed [1]	Chromatography	Detection System	Chromatographic Conditions and Method Validation Results	Reference(s)
i) (a) Rosmarinus officinalis, Origanum majorana, Origanum vulgare; (b) Ocimum basilicum, Mentha spicata, Thymus vulgaris; (c) Mentha × piperita, Thymus × citriodorus ii) (a) Fresh; (b) Dried; (c) Organic-dried	C5, **C16**, C36, C40, C50, C55, C58, C59, C61, C62, **C63**, C64, C66, C69, **C74**, C75, C78, **C79**	ultra-high-performance liquid chromatography (UHPLC)	DAD	Column: Acquity 'ethylene e bridged hybrid (BEH C18 (50 mm × 2.1 mm i.d., 1.7 μm p.s.) with an Acquity UHPLC BEH C18 VanGuard pre-column (5 mm × 2.1 mm i.d., 1.7 μm p.s.) (T): 20 °C Eluents: (A) 0.1% acetic acid in H$_2$O; (B) 0.1% Acetic acid in MeCN Run (t): 30 min LOD: 0.01–0.38 μg/mL, LOQ: 0.04–1.14 μg/mL, (R^2): ≥0.9990	[14]
i) 3Mentha ssp. ii) Dried and powdered leaves	C1, C4, C7, C21, **C28**, C46, C48, C57, C58, C59, **C63**, **C64**, C66, C70	HPLC	DAD	Column: Grace^{TM} Alltech^{TM} Alltima^{TM} C18 (250 mm × 4.6 mm i.d., 5.0 μm p.s.) (T): 40 °C Eluents: (A) MeCN: H$_2$O: formic acid (19:80:1); (B) MeCN: MeOH: formic acid (59:40:1) Run (t): 45 min	[7]
i) (a) 11 species of Mentha, (b) 2 Mixtures of Mentha species ii) (a) Plant material; (b) Finished Pharmaceutical products (2 Manufactures)	C1, C3, C10, C17, C21, C22, C28, C32, C46, C57, C63	two-dimensional micro-thin layer chromatography (2D-mTLC)	UV	Plate: HPTLC CNF 254 (10 cm × 10 cm, in 5 cm × 5 cm squares) Derivatization reagent: Naturstoff reagent 1st Condition: Non-aqueous eluent: 40% propan-2-ol in n-heptane; Aqueous eluent: 30% MeCN 2nd Condition: Non-aqueous eluent: 80% AcOEt in n-heptane; Aqueous eluent: 50% aqueous MeOH Sample quantity: 5 μL Conditioning: 20–30 min	[48]
i) Thymus vulgaris; Salvia officinalis ii) Aerial parts	C17, C19, C21, C22, C40, C45, C46, C57, C63, C64, C74	TLC	UV	Plate: Pre-coated silica gel TLC plates Si60 F254 Derivatization reagent: natural products-polyethylene glycol reagent (NP/PEG); 2,2-diphenyl-1-picrylhydrazyl radical (DPPH• in 0.2% in MeOH); Wavelength: 366 nm Eluents: For flavonoid aglycones: toluene: diethyl ether: acetic acid (60:40:10); For flavonoid glycosides: AcOEt: acetic acid: formic acid: H$_2$O (100:11:11:26); For phenolic acids: chloroform: ethyl acetate: acetone: formic acid (40:30:20:10)	[58]
		HPLC	DAD; MS in positive ion mode	Column: Zorbax Eclipse Plus PAH C18 (100 mm × 2.1 mm i.d. × 1.8 μm p.s.) Eluents: (A) 0.1% formic acid in H$_2$O; (B) 0.1% formic acid in MeCN Run (t): 30 min	

Table 2. Cont.

i) Lamiaceae Species ii) Plant Part	Polyphenols Analysed [1]	Chromatography	Detection System	Chromatographic Conditions and Method Validation Results	Reference(s)
i) Origanum vulgare, Ocimum basilicum, Rosmarinus officinalis, Origanum majorana, Thymus vulgaris, Satureja hortensis ii) Commercial, dried, grounded leaves	C63, C81	HPLC	UV-DPPH•; electrospray ionization (ESI)-MS in negative and positive ion mode	Column: Synergi Max-RP C12 (250 mm × 4.6 mm i.d., 4.0 μm p.s.) (T): 25 °C Eluents: (A) 0.05% TFA in H_2O; (B) 60% MeCN in MeOH Run (t): 60 min	[2]
i) Rosmarinus officinalis; Origanum vulgare; Salvia officinalis; Thymus vulgaris; Origanum vulgare ii) Leaves, or herbal mix, or as ingredients in chimichurri sauce	C63, C76 *, C78, C79 *, C80 *	HPLC	PDA	Column: Kinetex Polar C18 (250 mm × 4.6 mm i.d., 2.6 μm p.s.) (T): 55 °C Eluents: (A) 0.1% acetic acid in H_2O; (B) 0.1% acetic acid in MeCN Run (t): 10 min LOD: 0.25 μg/mL, LOQ: 1.0 μg/mL, (R^2): ≥0.9998	[39]
		UHPLC	MS in negative ion mode	Column: Acquity UHPLC BEH C18 (50 mm × 2.1 mm i.d., 1.7 μm p.s.) (T): 55 °C Eluents: (A) 0.1% acetic acid in H_2O; (B) 0.1% acetic acid in MeCN Run (t): 10 min	
i) 3 species of Salvia ii) Aerial parts, dried	Tentative identification only	LC	DAD-ESI-MS in positive ion mode	Column: Phenomenex Superspher 100 RP C18 (125 mm × 4.6 mm i.d. × 4.0 μm p.s.) (T): 40 °C Eluents: (A) 2.5% acetic acid in H_2O; (B) MeOH Run (t): 30 min	[46]
i) Satureja montana ssp. kitaibelii ii) Aerial parts of wild plant, air-dried	C17, C40, C46, C57, C59, C64, C69, C73	HPLC	DAD-ESI-time-of-flight (TOF)-MS	Column: Agilent Poroshell 120 C18 endcapped (100 mm × 4.6 mm i.d., 2.7 μm p.s.) (T): 25 °C Eluents: (A) 1% acetic acid in H_2O; (B) MeCN Run (t): 36 min LOD: 0.187–2.471 μg/mL, LOQ: 0.623–8.238 μg/mL, (R^2): ≥0.9983	[56]
i) Sicilian Origanum vulgare ssp. hirtum, Rosmarinus officinalis, Thymus capitatus L. ii) Dried-aerial parts	C1, C9, C13, C14 *, C17, C21, C57, C63, C74 *, C75 *, C78 *, C79 *, C80 *	HPLC	PDA/ESI-MS in positive and negative ion mode	Column: Phenomenex Luna C18 endcapped (250 mm × 4.6 mm i.d., 5.0 μm p.s.) (T): 25 °C Eluents: (A) 1% formic acid in H_2O; (B) MeCN Run (t): 64 min	[61–63]
i) Dried-aerial parts, flowering season, samples from various sites		GC	flame ionization detector (FID)/MS	Column: SPB-5 capillary (15 m length × 0.1 mm i.d. × 0.15 μm thickness) Injection: Split ratio (1:200)Oven (T): 60 °C for 1 min, linearly rising from 60 to 280 °C with a rate of 10 °C/min, 280 °C for 1 min	

Table 2. Cont.

i) Lamiaceae Species ii) Plant Part	Polyphenols Analysed [1]	Chromatography	Detection System	Chromatographic Conditions and Method Validation Results	Reference(s)
i) (a) Origanum majorana; (b) Mentha pulegium; (c) Lavandula officinalis ii) (a) Leaves and aerial parts; (b) Flowers; (c) Leaves, dried, milled	C1, C17, C34, C40, C46, C48, C51, C52, C57, C58, C59, C60, C63, C66, C67, C68, C69	UHPLC	DAD; ESI-tandem mass spectrometry (MS/MS) in negative ion and multiple reaction monitoring (MRM) mode	Column: Acquity UHPLC BEH C18 (100 mm × 2.1 mm i.d., 1.7 μm p.s.) (T): 30 °C Eluents: (A) 1% formic acid in H_2O; (B) 1% formic acid in MeOH Run (t): 12 min LOD: 0.02–5.52 ng/mL, LOQ: 0.06–18.20 ng/mL, linear regression (r): ≥0.9988	[68]
i) Thymus × citriodorus ii) Mixture of leaves and stems, dried	C2, C8, C19, C20, C20 *, C22, C23 *, C24, C63	RP-HPLC	DAD; ESI-MS and multi-stage mass spectrometry (MS^n) in negative ion mode; nuclear magnetic resonance (NMR)	Column: Nucleosil C18 endcapped (250 mm × 4.0 mm i.d., 5.0 μm p.s.) (T): 30 °C Eluents: (A) 0.1% formic acid in H_2O; (B) MeCN Run (t): 30 min LOD: 1.0–12.4 μg/mL, LOQ: 3.0–37.7 μg/mL, (R^2): ≥0.9984	[59]
i) Origanum majorana ii) Commercially produced, dried/grounded	C17, C22, C37, C40, C62 *, C63, C66	LC	ESI-MS/MS in negative ion mode; ^1H NMR	Column: Atlantis T3 C18 (100 mm × 2.1 mm i.d. × 3 μm p.s.) (T): 40 °C Eluents: (A) 0.5% formic acid in H_2O; (B) 0.5% formic acid in (MeCN: MeOH, 50:50) Run (t): 26 min	[50]
i) Rosmarinus officinalis; Origanum vulgare; Origanum majorana; Thymus vulgaris ii) Dried, grounded	C34, C36, C40, C57, C58, C59, C63, C67, C64, C69, C70	LC	PDA; ESI-linear ion trap quadrupole (LTQ)-Orbitrap-MS in negative ion mode	Column: Atlantis T3 RP C18 (100 mm × 2.1 mm i.d., 3 μm p.s.) (T): 25 °C Eluents: (A) 0.1% formic acid in H_2O; (B) 0.1% formic acid in MeCN Run (t): 36 min LOD: 1.7×10^{-3}–8.9×10^{-3} μg/g DW	[11,51]
i) Rosmarinus officinalis ii) Leaves from 20 different geographical zones	C6, C22, C25, C26, C27 *, C35 *, C37, C63, C67, C78, C79	HPLC	ESI-QTOF-MS and MS/MS in negative ion mode	Column: Zorbax Eclipse Plus C18 (150 mm × 4.6 mm i.d., 1.8 μm p.s.) (T): ≈20–25 °C Eluents: (A) 0.1% formic acid in H_2O; (B) MeCN Run (t): 30 min LOD: 0.014–0.24 μg/mL, LOQ: 0.04–0.8 μg/mL, (R^2): ≥0.9803	[67]
i) Mentha pulegium, Origanum majorana ii) Aerial parts	C13, C17, C21, C37, C54	RP-UHPLC	ESI-QTOF-MS and MS/MS in negative ion mode	Column: Zorbax Eclipse Plus C18 (150 mm × 4.6 mm i.d., 1.8 μm p.s.) (T): 25 °C Eluents: (A) 0.5% acetic acid in H_2O; (B) MeCN Run (t): 33 min	[52]
i) Mentha spicata ii) Commercial extract	C3, C31, C46, C63, C64, C65, C82, C83, C84	UHPLC	$ESI-MS^n$ in negative ion mode	Column: BlueOrchid C18 (50 mm × 2.0 mm i.d., 1.8 μm p.s.) (T): 30 °C Eluents: (A) 0.1% formic acid in H_2O; (B) 0.1% formic acid in MeCN Run (t): 20 min	[56]

Table 2. *Cont.*

i) *Lamiaceae* Species ii) Plant Part	Polyphenols Analysed [1]	Chromatography	Detection System	Chromatographic Conditions and Method Validation Results	Reference(s)
i) *Thymus serpyllum* ii) Whole-dried	C7, C21, C22 *, C39, C46, C57, C63, C64, C66, C69, C74, C75	RP-LC	DAD-ESI-MS/MS FID; mass selective detector (MSD);	Column: Phenomenex RP C18 (250 mm × 4.6 mm i.d. × 5.0 μm p.s.) (T): 25 °C Flow rate: 0.7 mL/min Eluents: (A) 1% formic acid in H_2O; (B) (MeCN/Solvent A) (60:40) Run (t): 106 min	[64]
		GC	mass spectrometry-olfactometry (MS-O)	Column: DB-Wax column (30 m length × 0.25 mm i.d. × 0.5 μm thickness) Injection: Pulsed splitless (40 psi; 0.5 min) Injector (T): 270 °C FID (T): 280 °C Oven (T): 250 °C for 10 min (50–250 °C with a rate of 4 °C/min)	
i) *Rosmarinus officinalis* ii) Branded extract rich in carnosic acid	C4, C18, C46, C57, C63, C76, C77, C78, C79	UHPLC	ESI-MS^n in negative ion mode	Column: XSelect HSS T3 C18 (50 mm × 2.1 mm i.d., 2.5 μm p.s.) (T): 30 °C Eluents: (A) 0.1% formic acid in H_2O; (B) 0.1% formic acid in MeCN Run (t): 35 min	[53]
		HPLC	PDA	Column: Phenomenex Luna C18 endcapped (250 mm × 4.6 mm i.d., 5.0 μm p.s.) (T): 30 °C Eluents: (A) 2.5% acetic acid in H_2O; (B) MeCN Run (t): 34 min Fraction collection (major peaks): Column: Phenomenex Luna 10 μm C18 (250 mm × 15 mm) Eluents: Similar to HPLC Run (t): Similar to HPLC	[65]
i) *Mentha australis* R. Br ii) Fresh leaves and stems	C1, C4, C5, C11 *, C17, C57, C63, C64	LC	Heated electrospay ionization (HESI)/atmospheric pressure chemical ionization	Similar conditions with HPLC. LOD: 0.25 ng	
		LC	(APCI)-MS/MS positive and negative ion mode; NMR HESI/APCI-high resolution mass spectrometry (HRMS) in positive and negative ion mode	Column: Phenomenex Synergi Hydro-RP C18 (250 mm × 1.0 mm i.d., 4.0 μm p.s.) (T): 45 °C Eluents: (A) 5 mM ammonium formate in H_2O (pH 7.4) (B) 5 mM ammonium formate in 90% aqueous MeOH (pH 7.4) Run (t): 19 min LOD: 0.625 ng	

Table 2. *Cont.*

i) *Lamiaceae* Species ii) Plant Part	Polyphenols Analysed [1]	Chromatography	Detection System	Chromatographic Conditions and Method Validation Results	Reference(s)
i) 6 *Ocimum* ssp. ii) Leaves, dried, grounded	C17, C21, C40, C43, **C46**, C48, C49, C55, C57, C59, C60, **C63**, C64, C66, C69	UHPLC	ESI-hybrid linear ion trap (QqQLIT) in negative ion mode	Column: Acquity UHPLC BEH C18 (50 mm × 2.1 mm i.d., 1.7 μm p.s.) (T): 50 °C Eluents: (A) 0.1% formic acid in H_2O; (B) 0.1% formic acid in MeCN Run (t): 13 min LOD: 0.041–0.357 ng/mL, LOQ: 0.124–1.082 ng/mL	[54]
i) 10 *Salvia* species ii) Plant material, dried	C17, **C20**, C21, C23, C29, C30, C42, C45, C47, C53, **C57**, **C63**, C71, C78, C79	HPLC	UV-DPPH•-MS PDA	Column: Discovery HS C18 (250 mm × 4.6 mm i.d., 5.0 μm p.s.) Flow rate: 0.8 mL/min Injection Volume: 20 μL Eluents: (A) 0.1% formic acid in H_2O; (B) MeOH Run (t): 60 min	[3]
		UHPLC	ESI-QTOF, triple quadrupole-spectrometer (TQ-S) in negative mode	Column: Acquity UHPLC BEH C18 (100 mm × 2.1 mm i.d., 1.7 μm p.s.) (T): 40 °C Eluents: (A) 0.1% formic acid in H_2O; (B) MeCN Run (t): 11 min LOD: 1.67–13.39 μg/mL, LOQ: 5.56–44.65 μg/mL	

[1] The reference analytical standards employed in each research. Note: The letter C followed by numbers correspond to the chemical structures and names that are given in Figure S1 (a, b, c, d, and e). The 'bold' compounds represent the most abundant polyphenols in the species analysed. The 'bold' compounds followed by *, represent the most abundant polyphenols that were tentatively quantified in the species analysed.

Many studies had been published in the past concerning the elucidation of phenolic profiles of various *Lamiaceae* herbs and spices through HPLC or RP-HPLC [76–81]. Nonetheless, more recent studies have also employed these techniques for the same purpose. HPLC analysis with a UV-diode array detector (DAD) was used by Chan, Gan, and Corke [13] for the examination of free (unbound) and bound phenolics (Table 2) in extracts of wild marjoram or oregano (*Origanum vulgare*) and additional herbs and spices [13], considering that bound phenolics encompass a considerable amount of the total phenolics in a matrix [82]. RP-HPLC coupled to UV/Vis-DAD was employed in the research of Žugić et al. [46] and elucidated 12 phenolic compounds in various plants, including European pennyroyal mint (*Mentha pulegium*) and hairless cat-mint (*Nepeta nuda*) (Table 2) [46]. Recently, Skendi, Irakli, and Chatzopoulou [50] developed a simple and reliable RP-HPLC technique with satisfactory sensitivity, reproducibility, accuracy and precision (Table 2) for the qualification and quantification of 24 phenolic compounds in botanicals of the *Lamiaceae* family, by optimizing the mobile phase composition and improving the separation of chromatographic peaks. The limit of detection (LOD) and limit of quantification (LOQ) were sufficiently low for identifying and qualifying low quantities of phenolic compounds, whereas the linearity was also good ($R^2 \geq 0.9961$). The phenolic content of the methanolic and aqueous extracts of the studied species declined as follows: Greek oregano (*Origanum vulgare* ssp. *hirtum*) > conehead thyme (*Thymus capitatus*) > winter savory (*Satureja thymbra*) > *Melissa officinalis* > rosemary (*Rosmarinus officinalis*) [50]. An HPLC method with UV/Vis detector was also developed and validated by Arceusz and Wesolowski [49] to evaluate the quality consistency of *Melissa officinalis*. Commercial herbs, while the optimized HPLC method was employed for the separation, identification and quantitation of six phenolic acids detected in this herb (Table 2) [49].

In the recent years, on-line HPLC-2,2, diphenyl-1-picrylhydrazyl radical (DPPH•) assay had been additionally used to effectively screen for the fast identification of antioxidant compounds from herbal extracts [83,84]. A simultaneous detection and quantification of compounds in complex plant matrices with high antioxidant potentials have also been investigated through on-line HPLC-UV-DPPH• analysis [2,3]. This technique was used by Damašius et al. [2] on extracts from different species of *Lamiaceae* family (Table 2). The authors concluded, that a strong correlation was found between antioxidant levels using the DPPH• bulk assay with that measured by the summed peak area attained through the on-line HPLC/UV/DPPH•. One phenolic acid, i.e., lithospermic acid B, was identified for the first time in marjoram (*Origanum majorana*), savory (*Satureja hortensis*) and thyme (*Thymus vulgaris*) (Table S1) [2]. The same technique was used adapted by Šulniūtė, Pukalskas, and Venskutonis [3] to identify rapidly the compounds with antioxidant potential in the extracts of different sage species (*Salvia* spp.) [3].

With advances in chromatography technologies in the past decade, ultra-high performance liquid chromatography (UHPLC) has enabled rapid separation of phenolics with much reduced time and cost [52]. UHPLC or UPLC is a chromatographic technique that is commercially available since 2004, and its applications have been rising steadily also for the qualification and quantification of the major phenolic compounds of several *Lamiaceae* herbs and spices [85–87]. The capability of higher pressure that ranges up to 15,000 psi (1035 bar) [86,88] and smaller particle size (potentially lower than 2 μm) [55,86,88], result in more rapid [55,86,88,89] effective [86], and sensitive separation of analytes [88]. Besides HPLC and UHPLC, there are other chromatography-based separation techniques that have been employed for phenolic profile characterization, such as CE and TLC. These techniques, in particular CE, can also be hyphenated to MS for acquisition of structural data [72].

TLC is a rapid and easy-to-use technique that can be employed for initial identification of phenolics in various extracts [82]. Even if the popularity of TLC has decreased as a result of the advance of column chromatography, it remains an essential tool in the research of polyphenols in natural extracts [58]. TLC and HPLC with DAD detection system were used by Fatiha et al. [57] in order to diminish the probability of misidentification, throughout elucidation of the phenolic profiles of extracts of mint subspecies (*Mentha* spp.) (Table 2). TLC and HPLC analysis revealed similar phenolic compounds (caffeic acid, rosmarinic acids, and diosmin) as well as their derivatives were identified with both techniques in all extracts [57]. Jesionek, Majer-Dziedzic, and Choma [58] optimized a TLC technique and separated 10 typical phenolic constituents from five plant species extracts, including *Thymus vulgaris* and common sage (*Salvia officinalis*) (Table 2) [58]. In parallel, a TLC-DPPH• assay was used to define the antioxidant capacity of the extracts, and liquid chromatography coupled to mass spectrometry (LC-MS) as a confirmation tool of the occurrence of the targeted phenolics. The separation of polyphenols on TLC is typically accomplished with silica gel and AcOEt:acetic acid:formic acid:water (100:11:11:26, v/v) as a mobile phase. Nonetheless, seven different mobile phases were used to optimize the separation of polyphenols, while two novel were ultimately established and utilized. The optimized eluent system enabled the good separation of phenolic compounds and correspondingly their clear detection. Apigenin 7-*O*-glucoside was the only phenolic compound that did not display any antioxidant capacity through TLC–DPPH• assay, while most likely, the low concentration of the four additional phenolic constituents identified through LC-MS was the factor that restricted their detection through TLC [58].

Regardless of its low resolution [82], TLC represents a valuable technique as it can be easily setup for 2-D chromatography, whereas post-separation derivatization process can deliver further analyte selectivity [72]. Two-dimensional (2D) LC or LCxLC offers enhanced resolution of complex matrices and is becoming extensively utilized due to the improved characterization of compounds with respect to one-dimensional liquid chromatography [90]. In some cases, analysis of phenolic in herbs and spices by conventional chromatographic techniques is challenging especially when key components cannot be effectively resolved, indicating the demand of effective multi-dimensional separation techniques. An LC × LC system is constituted in most of the cases by two different separation columns which results in the efficient qualification and quantification of compounds. Subsequently, improved MS analysis can be achieved as the matrix-associated ionization suppression is minimized [91]. In the research of Hawryl et al. [48], a micro-2D-TLC method with cyanopropyl layers led to the separation of phenolic fractions from several mint species (*Mentha* sp.) extracts (Table 2). The 2D-TLC data indicated the presence of rutin, narirutin, rosmarinic acid, isorhoifolin, diosmin, and naringenin in all the *Mentha* sp. extracts. Initially, the technique was optimized through the utilization of different concentrations of MeCN and H_2O. Subsequently, the eluents with the higher selectivity were used to optimize the 2D systems through the development of R_f (Retention factors) on the TLC plates, for both normal and reversed phases. It was noted that the 2D-TLC technique was highly sensitive, time efficient, and required low volumes of eluent and sample [48].

Separation and analysis of polyphenols in herbs and spices by CE involves separation based on the electrophoretic mobilities of a solution that consists of electrically charged species, in small-diameter capillaries [92] it is recognized as being effective in phenolic characterization, offering practical operation, rapid analysis, low consumption of solvent, and low cost. This method represents a valuable alternative to HPLC in the separation of closely associated phenolics, but its major drawbacks are its lower reproducibility and sensitivity as compared to HPLC [93]. Maher et al. [47] used an optimized CE with DAD to identify luteolin and apigenin in *Thymus vulgaris* and an additional herb extract (Table 2). The technique was optimized in terms of voltage, capillary temperature, applied pressure, detection wavelength, as well as pH and buffer, and MeOH concentration. The principal advantages of the CE technique were its selectivity for the analytes, deprived from interferences from other compounds, its short analysis time (less than 35 min) and the ease of use. In parallel, it was characterized as sensitive, accurate and precise [47].

4. Hyphenated Chromatographic Techniques

Over the last two decades, hyphenation of chromatographic and spectroscopic techniques has gained considerable esteem in the analysis of complex biological matrices [94]. Mass spectrometer coupled to LC or GC constitutes the most widely used hyphenated analytical methods in the analysis of food components [95]. The basic principle of MS is the generation of ions in gas phase from either organic or inorganic compounds, the separation of ions based on their mass-to-charge ratio (m/z) and the qualitative and quantitative detection of the components through their respective m/z and abundance [96]. For the molecules that do not ionise readily, atmospheric pressure chemical ionization (APCI) to assist ionization has been used in the LC-MS methods [65,97–99].

LC-MS [3,100–102] and LC-MS/MS [103–105] have been widely used for the characterization of the phenolic profiles of various herbs and spices. LC-DAD-MS was used by Atwi et al. [66] to analyse three sage (*Salvia*) species (Table 2), native in Crete Island (Greece), in AcOEt and *n*-butanol extracts. As the chromatographic analysis revealed, the different species had a high phenolic content, predominantly in flavones, while a restricted amount of phenylpropanoids was also present. Additionally, Greek sage (*Salvia fruticosa*) *n*-butanol extracts showed the highest antioxidant capacity [66]. In addition, Milevskaya et al. [70] used LC-DAD-MS analysis to qualify and identify the extracted phenolic compounds from 4 Lamiaceae herbs, namely *Salvia officinalis* L., creeping thyme (*Thymus serpyllum*), *Origanum vulgare*, and *Melissa officinalis*) by utilizing different extraction processes (Table 2). Subcritical extraction resulted in the highest extractability of phenolics, while *Origanum vulgare* exhibited the maximum content in some of them. Nonetheless, the researchers also suggested that the comparison of the UV spectra and retention times of analytes and standards is not adequate for qualifying phenolics in medicinal plants, while the supplementary use of MS could provide higher reliability to the process [70]. Tuttolomondo et al. [61] applied HPLC-PDA/ESI-MS on the analysis of phytochemicals in 57 wild Sicilian oregano (*Origanum vulgare* ssp. *hirtum*) samples (Table 2), where 13 polyphenol derivatives (flavanones, flavones, organic acids) were quantified and showed that flavanones were more abundant that the flavones [61]. In the subsequent studies by the same research group on wild Sicilian *Rosmarinus officinalis* L. [62] and wild Sicilian thyme (*Thymus capitatus* L.) [63], eighteen compounds (flavones, diterpenes, organic acids) and fifteen flavonoid derivatives were identified in the respective Lamiaecea species examined [62,63].

LC-MS/MS was used by Sonmezdag, Kelebek, and Selli [64] for the characterization of the phenolic compounds of *Thymus serpyllum* (Table 2), after aqueous-alcoholic extraction, where 18 phenolic compounds were identified and quantified; of which 10 of the 18 compounds were reported for the first time in this species (Table S1). Except for luteolin 7-*O*-glucoside that was the predominant compound of the phenolic fraction, luteolin and rosmarinic acid were also detected in considerable quantities [64]. In another study, Hossain et al. [60] employed LC-ESI-MS/MS (Table 2) to qualitatively and quantitatively examine antioxidant-guided polyphenol rich fractions of *Origanum majorana*, following flash chromatography (FC). The study revealed that rosmarinic acid, confirmed with ^{1}H nuclear magnetic resonance (NMR) data, mainly attributed to the antioxidant activity of *Origanum majorana* [60]. FC constitutes on of the simplest methodologies of maximizing the quantities and purity of natural active isolates, for their supplementary structural interpretation through NMR spectroscopy. Regardless its lower resolution compared to other techniques, FC has the benefits of being simple and inexpensive [106].

NMR spectroscopy is often used as a confirmatory tool in the identification of polyphenols [19]. NMR analysis is essential to establish the configuration of new molecules that have been reported for the first time, by measuring the total biochemical composition of a matrix [18,72]. However, the limiting factor for elucidation of chemical structures through NMR is the requirement of high quantities of the substances with excellent purity [72]. Particularly, ^{1}H NMR spectroscopy can deliver rapid, direct and without interferences profiling of polyphenols [82]. A combination of HPLC-DAD ESI-MS, MS^n and 2D-NMR (^{1}H, ^{13}C) analysis [59] were employed in profiling phenolic compounds of lemon thyme (*Thymus* x *citriodorus*) ethanolic extracts (Table 2). The in-house validation of this combined method gave rise to sufficient results of linearity (adjusted R^2 values ~0.999), instrumental and technique precision as well as accuracy, whereas LOD and LOQ values revealed an adequate sensitivity for all used phenolic standards. Among the 13 identified phenolics in *Thymus* x *citriodorus*, the major compound was rosmarinic acid. However, luteolin-7-*O*-glucuronide was also detected in high quantities for the first time in thyme (*Thymus*) species (Table S1), whereas other novel compounds were also present (Table S1) [59]. Several studies have demonstrated the application of UHPLC-MS/MS for phenolic profiling of herbal samples [55,64,66,68] which is deliberated as advanced, sensitive, reproducible, rapid and with high resolution technique [68]. For instance, Mena et al. [53] have used UHPLC-ESI-MS^n with a total run time of 35 min, for the phenolic compositional analysis of a branded extract of *Rosmarinus officinalis* (Table 2), where 57 compounds were identified and quantified, and of which 14 polyphenols were detected for the first time (Table S1) in this species [53]. In another UHPLC-ESI-MS^n study of methanolic extracts of dried *Mentha spicata* L., by Cirlini et al. [56], its (poly) phenolic profile was fully elucidated (Table 2), and 66 molecules were identified, whereas 53 of them were semi-quantified in a shorter time, equal to 20 min [56]. Compared to the conventional LC systems, UHPLC based separation methods are five to 10-fold faster with peak resolutions retained [89] or enhanced [55,88,89] whereas they result in lower limits of detection and reduced solvent consumption [14]. The benefits of these techniques stem from the used analytical columns, with particle size <2.0 μm, which lead to considerable reduction of back-pressure [14]. Polyphenolic profiles of *Lamiaceae* species, namely *Origanum majorana*, *Mentha pulegium* and lavender (*Lavandula officinalis*) were also scrutinised by Çelik et al. [68] on MAE 60% MeOH extracts (Table 2). The authors optimized and validated the UHPLC-DAD-ESI-MS/MS method that had a total run time of 12 min per sample. A total number of 18 polyphenols was identified in the samples and the technique exhibited good reproducibility (recoveries equal to 92–109%) and linearity (r $\geq$ 0.9988), whereas LOD and LOQ values of polyphenols were diminished to 0.02 ng/mL and 0.06 ng/mL, respectively. The advantages of this method over HPLC are attributed to the reduction of analysis time and its applicability to a greater number of polyphenolic compounds [68]. Oliveira et al. [14] developed and validated an UHPLC-DAD method (Table 2) that enabled for the first time the simultaneous quantification of 19 phenolic compounds in 21 fresh and dried (organic and non-organic) aromatic plants, most of them belonging to *Lamiaceae* family. This technique was capable of identifying and quantifying phenolic compounds at a concentration <0.15 μg/mL, apart from carnosol and carnosic acid, in a relatively short run time (30 min), whereas it was direct, sensitive, with good precision, accuracy and linearity. It was further revealed, among the different aromatic plants, *Thymus vulgaris* displayed the highest range of different phenolics [14].

Even if reduced particle size leads to increased column efficiency and analysis time, it also results in increased back-pressures. Fused-core technology is considered as a way of archiving both the benefits of small particles and the existing pressures with an HPLC system, consisting of 1.7 μm solid silica bead surrounded by a 0.5 μm porous shell, while deriving a particle size equal to 2.7 μm. One benefit of the fused-core columns is that for a certain column length, it does not involve the comparatively high pressures that are essential by a column packed with 1.7 μm material. Nonetheless, the overall column efficiency is reduced by 20% in comparison to a 1.7 μm packed bed [107]. Zabot et al. [69] employed UHPLC-MS to confirm the identified phenolic terpenes, while developing and validating a rapid HPLC-PDA technique through a fused-core column for their analysis in *Rosmarinus officinalis* (Table 2). Several chromatographic parameters were optimized (column temperature, gradient and flow rate, re-equilibration period), and the validated technique had the ability to detect and quantify the principal non-volatile constituents of *Rosmarinus officinalis* (carnosol, rosmanol, carnosic acid, rosmarinic acid, methyl carnosate) in low amounts of 0.25 μg/mL and 1 μg/mL, respectively. The analysis had a short total run time of 10 min and was shown to be convenient in use, selective, robust and reliable [69].

Liquid chromatography coupled to various mass spectrometers such as TOF, and Orbitrap attracting considerable interest the last years [108], rendering high resolution mass spectroscopy (HRMS) as a powerful structural elucidation tool [109]. The contemporary hybrid mass analysers, such as Q-TOFs and Q-Orbitraps, have led to remarkable technological developments in facilitating specific ion fragmentation and expedite data mining and thereby increase the potential for the identification of unknown compounds [110]. Except for providing improved specificity compared to conventional MS techniques, HRMS techniques correspondingly facilitate software expedite data mining. Even if reference standards are essential for conformation of identity, when they are absent, these methods have the capacity to tentatively or fully identify the unknown compounds [55,111] based on UV absorption, MS spectra and information in the literature [55]. LTQ-Orbitrap-MS is the most advanced mass spectrometry technique that allows rapid, accurate and sensitive structural elucidation of small molecules [11,112], without the effect of the relative ion abundance [112] and through MS, MS/MS as well as MS^n [11]. SPE followed by LC and coupled with ESI-LTQ-Orbitrap-MS [11] resulted in the identification of 52 polyphenolic compounds in several families of culinary herbs and spices including *Lamiaceae* (Table 2), despite the fact that standards were not employed in the analysis [11]. The principal compounds were also quantified through LC coupled to ESI-QqQ and multiple reaction monitoring (MRM mode with optimized conditions. Moreover, two polyphenols were identified for the first time in the examined *Lamiaceae* herbs (*Rosmarinus officinalis*, *Thymus vulgaris*, and *Origanum vulgare*) (Table S1) [11]. The same conditions were effectively used in the subsequent study of Vallverdú-Queralt et al. [51], for the analysis of the phenolic profile of five additional herbs, including *Origanum majorana* (Table 2), whereas 22 phenolics were identified in its extract [51]. Pandey et al. [54] developed an UHPLC coupled to QqQ$_{LIT}$-MS/MS in MRM mode, to investigate differences in the bioactive components, among them (poly) phenolic compounds, of leaf extracts of six basil (*Ocimum*) species (Table 2). The developed and validated technique was rapid, with a run time of 13 min, whereas it was characterized as sensitive, precise and reliable, according to the international standards. Among all the bioactive constituents and for almost all the examined *Ocimum* species, rosmarinic acid was the predominant phenolic constituent [54].

The accurate mass measurement of Q-TOF for precursor and product ions, constitute the factors of its wide application [113]. Extracts of leaves of 20 *Rosmarinus officinalis* plants originated from different areas of Serbia were analyzed by high performance liquid chromatography coupled to HPLC-ESI-Q-TOF-MS and MS/MS (Table 2) by Borrás-Linares et al. [67] Q-TOF mass analyzer resulted in the qualification and quantification of the 30 phenolic compounds (Table 2) and was established as an important detection system in phenolic characterization, offering mass accuracy and true isotopic spectral distribution in both MS and MS/MS [67]. HPLC–ESI–Q-TOF–MS was also employed in the research of López-Cobo et al. [55] and elucidated (Table 2) the phenolic profile of the wild growing winter savory (*Satureja montana* ssp. *kitaibelii*). In this study, a total of 44 phenolics were identified, of which 42 were identified for the first time in this species (Table S1) [55]. Šulniūtė, Pukalskas, and Venskutonis [3] examined 10 *Salvia* spp. species following SFE-CO$_2$ in EtOH and H$_2$O (Table 2). Subsequent analysis of this extract using UHPLC-Q-TOF and UHPLC-TQ-S was performed and showed that rosmarinic acid was the principal compound in various *Salvia* spp., mainly in ethanolic extracts. Additional polyphenols, i.e., apigenin glucuronide, caffeic and carnosic acids were identified and quantified for the first time in *Salvia* spp. (Table S1) [3]. Methanolic extracts of Tunisian *Mentha pulegium* and *Origanum majorana* were analyzed with UHPLC-Q-TOF-MS by Taamalli et al. [52]. The authors detected 85 metabolites from several chemical families, and among them were phenolic compounds, which were quantified spectrophotometrically based on the chromatographic peak areas. This study had shown higher amounts of polyphenols in *Mentha pulegium* extract than in *Origanum majorana*, and high-resolution mass spectra with accuracy of 5 ppm were delivered. According to the authors, this study enabled the characterization of several compounds belonging to different classes in a single run, and some of the compounds reported for the first time in this species (Table S1) [52].

Even if HRMS is effective in the detection of novel compounds, supplementary characterization is required for incontrovertible results, as for instance through the use of ^{1}H NMR and ^{13}C NMR analysis. Nonetheless, in most of the cases where new compounds are identified, adequate information is available to minimize the selection, attributed to a logical framework for extrapolation from identified compounds to the unidentified [114]. ^{1}H NMR, ^{13}C NMR including 2D NMR analyses in tandem with LC-MS/MS in MRM acquisition mode were utilized to validate the results of HPLC-PDA and LC-HRMS in the investigation of the phenolic profile of Australian mint (*Mentha australis* R. Br.) (Table 2). MRM mode is particularly specific and more sensitive compared to LC-HRMS. Therefore, it was employed to validate the chemical structures attained through LC-HRMS by scrutinizing the product ions of authentic standards and excluding the unwanted ions. Through this means, it enabled precision while relating to the standards. It was shown in this study that LC-HRMS delivered mass accuracy of less than 2 ppm. Except for rosmarinic acid and neoponcirin, gallic acid equivalent, narirutin, chlorogenic acid, and biochanin A were also identified as major compounds of *Mentha australis* R. Br., whereas two phenolic compounds were identified for the first time in the *Mentha genus* (Table S1) [65].

GC is also used in some cases for the quantification of phenolic compounds, in particular for volatiles [71]. Generally, fused silica capillaries of 30 m length and internal diameters of 25–32 µm, and a stationary phase particle size of 0.25 µm are used in GC. Flame ionization detector (FID) and MS are the commonly used detectors [23]. Although GC has been used particularly for identification and quantification of flavonoids and phenolic acids, the low volatility of phenolics is a deterrent factor requiring chemical derivatization (methylation) [44]. GC coupled to MS has been used in profiling phenolics in herbs and spices [23]. Two phenolic terpenes (thymol and carvacrol) were the main compounds in the essential oil of *Thymus serpyllum* as determined by GC-MS (Table 2). The volatile compounds were recovered, and their separation was carried out using a flame ionization detector (FID) and a mass-selective detector (MSD). Subsequently, the aroma extract dilution analysis of the extract was followed with GS-MS-O [64]. The GC-MS-O technique provides separation of the volatile compounds by odorous and non-odorous properties, based on their concentrations in the examined matrix [64]. In a separate study by Tuttolomondo et al., GC-FID and GC-MS analyses showed 81 compounds in the essential oils of wild Sicilian *Origanum vulgare* ssp. *hirtum*. obtained after hydrodistillation, and the principal compound in the extracted oils was the phenolic terpene thymol [61]. In the following studies by Napoli et al. [62] and Saija et al. [63], GC-FID and GC-MS analyses on wild Sicilian *Rosmarinus officinalis* L. and *Thymus capitatus* L. identified carvacrol as the major phenolic terpene in *Thymus capitatus* L. oils [63].

5. Conclusions

The promising results in last decades regarding the antioxidant and health-promoting properties of *Lamiaceae* merit the investigation of their active compounds, which are predominantly polyphenols. Advances in analytical technologies, such as hyphenated methods and multi-dimensional separation techniques, including UHPLC or LC × LC coupled to MS such as Orbitrap and Q-TOF, or NMR, have enabled the identification of several new polyphenols in *Lamiaceae* herbs, and in addition made it possible to quantify the low levels (nanograms) present in some matrices. Nonetheless, further development in analytical capabilities is required to distinguish the structural anomaly diversity of polyphenols and their metabolites (transformed by gut bacteria or enzymes) in a complex matrix.

Supplementary Materials: The following are available online at http://www.mdpi.com/2223-7747/7/2/25/s1, Figure S1 (a, b, c, d, e): The chemical structures of the analytical standards or the most abundant polyphenols in the analysed species, Table S1: (Poly) phenolic compounds identified for the first time in the literature cited in Table 2.

Acknowledgments: This work was supported by the Teagasc Walsh Fellowships Program (2016038), as a part of a doctorate research.

Author Contributions: Katerina Tzima, Nigel P. Brunton and Dilip K. Rai contributed to the conception and writing of the manuscript. Dilip K. Rai and Nigel P. Brunton proposed the topic provided ideas and contributed with editing and reviewing the manuscript. All authors have read and approved the final manuscript.

Conflicts of Interest: The authors declare no conflicts of interest.

Abbreviations

The following abbreviations are used in this Manuscript:

AcOEt	ethyl acetate
APCI	atmospheric pressure chemical ionization
ASE	accelerated solvent extraction
BHA	butylated hydroxyanisole
BHT	butylated hydroxytoluene

CE	capillary electrophoresis
CO_2	carbon dioxide
DAD	diode array
DPPH•	2,2-diphenyl-1-picrylhydrazyl radical
ESI	electrospray ionization
EtOH	ethanol
FC	flash chromatography
FID	flame ionization detection
GC	gas chromatography
hr	hours
H_2O	water
HCl	hydrochloric acid
HESI	heated electrospray ionization
HPLC	high-performance liquid chromatography
HRMS	high resolution mass spectrometry
i.d.	internal diameter
LC	liquid chromatography
LOD	limit of detection
LOQ	limit of quantification
LPSE	low pressure solvent extraction
LTQ	linear ion trap quadrupole
QqQLIT	hybrid linear ion trap
MAE	microwave assisted extraction
ME	matrix effects
MeCN	acetonitrile
MeOH	methanol
MRM	multiple reaction monitoring
MS	mass spectrometry
MS^n	multi-stage mass spectrometry
MS/MS	tandem mass spectrometry
MSD	mass selective detector
MS-O	mass spectrometry-olfactometry
m/zn/a	mass-to-charge ratio / not available
NaOH	sodium hydroxide
N_2	Nitrogen
NMR	nuclear magnetic resonance
NP/PEG	natural products-polyethylene glycol reagent
O_2	Oxygen
P & T	purge and trap
PDA	photodiode array
p.s.	particle size
r	linear regression
R^2	correlation/determination coefficient
RP	reversed phase
RSC	radical scavenging capacity
SFE	supercritical fluid extraction
SIM	selected ion monitoring mode
SPE	solid phase extraction
mTLC	micro-thin layer chromatography
T	temperature
t	Time
TFA	trifluoroacetic acid

TLC	thin layer chromatography
TOF	time-of-flight
TQS	triple quadrupole spectrometer
UAE	ultrasound-assisted extraction
UHPLC	ultra-high performance liquid chromatography
UV	ultraviolet
Vis	visible
v/v	volume/volume
w/v	weight/volume
2D	two dimensional

References

1. Hinneburg, I.; Dorman, H.D.; Hiltunen, R. Antioxidant activities of extracts selected culinary herbs and spices. *Food Chem.* **2006**, *97*, 122–129. [CrossRef]
2. Damašius, J.; Venskutonis, P.; Kaškoniene, V.; Maruška, A. Fast screening of the main phenolic acids with antioxidant properties in common spices using on-line HPLC/UV/DPPH radical scavenging assay. *Anal. Methods* **2014**, *6*, 2774–2779. [CrossRef]
3. Šulniūtė, V.; Pukalskas, A.; Venskutonis, P.R. Phytochemical composition of fractions isolated from ten *Salvia* species by supercritical carbon dioxide and pressurized liquid extraction methods. *Food Chem.* **2017**, *224*, 37–47. [CrossRef] [PubMed]
4. Babovic, N.; Djilas, S.; Jadranin, M.; Vajs, V.; Ivanovic, J.; Petrovic, S.; Zizovic, I. Supercritical carbon dioxide extraction of antioxidant fractions from selected *Lamiaceae* herbs and their antioxidant capacity. *Innov. Food Sci. Emerg. Technol.* **2010**, *11*, 98–107. [CrossRef]
5. Klejdus, B.; Kováčik, J. Quantification of phenols in cinnamon: A special focus on "total phenols" and phenolic acids including DESI-Orbitrap MS detection. *Ind. Crop. Prod.* **2016**, *83*, 774–780. [CrossRef]
6. Leja, K.B.; Czaczyk, K. The industrial potential of herbs and spices—A mini review. *Acta Sci. Pol. Technol. Aliment.* **2016**, *15*, 353–365. [CrossRef] [PubMed]
7. Calucci, L.; Pinzino, C.; Zandomeneghi, M.; Capocchi, A.; Ghiringhelli, S.; Saviozzi, F.; Tozzi, S.; Galleschi, L. Effects of γ-irradiation on the free radical and antioxidant contents in nine aromatic herbs and spices. *J. Agric. Food Chem.* **2003**, *51*, 927–934. [CrossRef] [PubMed]
8. Aghakhani, F.; Kharazian, N.; Gooini, Z.L. Flavonoid Constituents of Phlomis (*Lamiaceae*) Species Using Liquid Chromatography Mass Spectrometry. *Phytochem. Anal.* **2017**, *29*, 180–195. [CrossRef] [PubMed]
9. Putnik, P.; Kovačević, D.B.; Penić, M.; Fegeš, M.; Dragović-Uzelac, V. Microwave-assisted extraction (MAE) of dalmatian sage leaves for the optimal yield of polyphenols: HPLC-DAD identification and quantification. *Food Anal. Methods* **2016**, *9*, 2385–2394. [CrossRef]
10. Pan, M.H.; Lai, C.S.; Ho, C.T. Anti-inflammatory activity of natural dietary flavonoids. *Food Funct.* **2010**, *1*, 15–31. [CrossRef] [PubMed]
11. Vallverdú-Queralt, A.; Regueiro, J.; Martínez-Huélamo, M.; Alvarenga, J.F.R.; Leal, L.N.; Lamuela-Raventos, R.M. A comprehensive study on the phenolic profile of widely used culinary herbs and spices: Rosemary, thyme, oregano, cinnamon, cumin and bay. *Food Chem.* **2014**, *154*, 299–307. [CrossRef] [PubMed]
12. Engel, R.; Szabó, K.; Abrankó, L.S.; Rendes, K.; Füzy, A.; Takács, T.N. Effect of Arbuscular Mycorrhizal Fungi on the Growth and Polyphenol Profile of Marjoram, Lemon Balm, and Marigold. *J. Agric. Food Chem.* **2016**, *64*, 3733–3742. [CrossRef] [PubMed]
13. Chan, C.L.; Gan, R.Y.; Corke, H. The phenolic composition and antioxidant capacity of soluble and bound extracts in selected dietary spices and medicinal herbs. *Int. J. Food Sci. Technol.* **2016**, *51*, 565–573. [CrossRef]
14. Oliveira, A.S.; Ribeiro-Santos, R.; Ramos, F.; Castilho, M.C.; Sanches-Silva, A. UHPLC-DAD Multi-Method for Determination of Phenolics in Aromatic Plants. *Food Anal. Methods* **2018**, *11*, 440–450. [CrossRef]
15. Quideau, S.; Deffieux, D.; Douat-Casassus, C.; Pouysegu, L. Plant polyphenols: Chemical properties, biological activities, and synthesis. *Angew. Chem. Int. Ed.* **2011**, *50*, 586–621. [CrossRef] [PubMed]
16. Boudet, A.M. Studies on quinic acid biosynthesis in *Quercus pedunculata* Ehrh. Seedlings. *Plant Cell Physiol.* **1980**, *21*, 785–792. [CrossRef]

17. Kalili, K.M.; de Villiers, A. Recent developments in the HPLC separation of phenolic compounds. *J. Sep. Sci.* **2011**, *34*, 854–876. [CrossRef] [PubMed]
18. Gad, H.A.; El-Ahmady, S.H.; Abou-Shoer, M.I.; Al-Azizi, M.M. Application of Chemometrics in Authentication of Herbal Medicines: A Review. *Phytochem. Anal.* **2013**, *24*, 1–24. [CrossRef] [PubMed]
19. Costa, D.C.; Costa, H.; Albuquerque, T.G.; Ramos, F.; Castilho, M.C.; Sanches-Silva, A. Advances in phenolic compounds analysis of aromatic plants and their potential applications. *Trends Food. Sci. Technol.* **2015**, *45*, 336–354. [CrossRef]
20. Popova, I.E.; Hall, C.; Kubátová, A. Determination of lignans in flaxseed using liquid chromatography with time-of-flight mass spectrometry. *J. Chromatogr. A* **2009**, *1216*, 217–229. [CrossRef] [PubMed]
21. Ganzera, M.; Sturm, S. Recent advances on HPLC/MS in medicinal plant analysis—An update covering 2011–2016. *J. Pharm. Biomed. Anal.* **2018**, *147*, 211–233. [CrossRef] [PubMed]
22. Tubaon, R.M.; Haddad, P.R.; Quirino, J.P. Sample Clean-up Strategies for ESI Mass Spectrometry Applications in Bottom-up Proteomics: Trends from 2012 to 2016. *Proteomics* **2017**, *17*. [CrossRef] [PubMed]
23. Khoddami, A.; Wilkes, M.A.; Roberts, T.H. Techniques for analysis of plant phenolic compounds. *Molecules* **2013**, *18*, 2328–2375. [CrossRef] [PubMed]
24. Dobiáš, P.; Pavlíková, P.; Adam, M.; Eisner, A.; Beňová, B.; Ventura, K. Comparison of pressurised fluid and ultrasonic extraction methods for analysis of plant antioxidants and their antioxidant capacity. *Cent. Eur. J. Chem.* **2010**, *8*, 87–95. [CrossRef]
25. Garcia-Salas, P.; Morales-Soto, A.; Segura-Carretero, A.; Fernández-Gutiérrez, A. Phenolic-compound-extraction systems for fruit and vegetable samples. *Molecules* **2010**, *15*, 8813–8826. [CrossRef] [PubMed]
26. Azmir, J.; Zaidul, I.S.M.; Rahman, M.M.; Sharif, K.M.; Mohamed, A.; Sahena, F.; Jahurul, M.H.A.; Ghafoor, K.; Norulaini, N.A.N.; Omar, A.K.M. Techniques for extraction of bioactive compounds from plant materials: A review. *J. Food Eng.* **2013**, *117*, 426–436. [CrossRef]
27. Alonso-Carrillo, N.; de los Ángeles Aguilar-Santamaría, M.; Vernon-Carter, E.J.; Jiménez-Alvarado, R.; Cruz-Sosa, F.; Román-Guerrero, A. Extraction of phenolic compounds from *Satureja macrostema* using microwave-ultrasound assisted and reflux methods and evaluation of their antioxidant activity and cytotoxicity. *Ind. Crop. Prod.* **2017**, *103*, 213–221. [CrossRef]
28. Santos, D.T.; Veggi, P.C.; Angela, M.; Meireles, A. Optimization and economic evaluation of pressurized liquid extraction of phenolic compounds from jabuticaba skins. *J. Food Eng.* **2012**, *108*, 444–452. [CrossRef]
29. Hossain, M.B.; Barry-Ryan, C.; Martin-Diana, A.B.; Brunton, N.P. Optimisation of accelerated solvent extraction of antioxidant compounds from rosemary (*Rosmarinus officinalis* L.), marjoram (*Origanum majorana* L.) and oregano (*Origanum vulgare* L.) using response surface methodology. *Food Chem.* **2011**, *126*, 339–346. [CrossRef]
30. Laroze, L.; Soto, C.; Zúñiga, M.L. Raspberry phenolic antioxidants extraction. *J. Biotechnol.* **2008**, *136*, 717–742. [CrossRef]
31. Cheok, C.Y.; Salman, H.A.K.; Sulaiman, R. Extraction and quantification of saponins: A review. *Food Res. Int.* **2014**, *5*, 16–40. [CrossRef]
32. Moraes, M.N.; Zabot, G.L.; Prado, J.M.; Meireles, M.A.A. Obtaining antioxidants from botanic matrices applying novel extraction techniques. *Food Public Health* **2013**, *3*, 195–214. [CrossRef]
33. Mojzer, E.B.; Hrnčič, M.K.; Škerget, M.; Knez, Ž.; Bren, U. Polyphenols: Extraction Methods, Antioxidative Action, Bioavailability and Anticarcinogenic Effects. *Molecules* **2016**, *21*, 901. [CrossRef] [PubMed]
34. Conidi, C.; Drioli, E.; Cassano, A. Membrane-based agro-food production processes for polyphenol separation, purification and concentration. *Curr. Opin. Food Sci.* **2017**, in press, Corrected Proof. [CrossRef]
35. Sadeghi, A.; Hakimzadeh, V.; Karimifar, B. Microwave Assisted Extraction of Bioactive Compounds from Food: A Review. *Int. J. Food Sci. Nutr. Eng.* **2017**, *7*, 19–27. [CrossRef]
36. Ekezie, F.G.C.; Sun, D.W.; Han, Z.; Cheng, J.H. Microwave-Assisted Food Processing Technologies for Enhancing Product Quality and Process Efficiency: A Review of Recent Developments. *Trends Food Sci. Technol.* **2017**, *67*, 58–69. [CrossRef]
37. Zhang, H.F.; Yang, X.H.; Wang, Y. Microwave assisted extraction of secondary metabolites from plants: Current status and future directions. *Trends Food Sci. Technol.* **2011**, *22*, 672–688. [CrossRef]

38. Aires, A. Phenolics in Foods: Extraction, Analysis and Measurements. In *Phenolic Compounds—Natural Sources, Importance and Applications*, 1st ed.; Soto-Hernandez, M., Palma-Tenango, M., Garcia-Mateos, M.R., Eds.; InTech: London, UK, 2017; pp. 61–88, ISBN 978-953-51-2958-5. Available online: https://www.intechopen.com/books/phenolic-compounds-natural-sources-importance-and-applications (accessed on 25 March 2018).
39. Herrero, M.; Cifuentes, A.; Ibañez, E. Sub- and supercritical fluid extraction of functional ingredients from different natural sources: Plants, food-by-products, algae and microalgae: A review. *Food Chem.* **2006**, *98*, 136–148. [CrossRef]
40. Barbosa-Pereira, L.; Pocheville, A.; Angulo, I.; Paseiro-Losada, P.; Cruz, J.M. Fractionation and Purification of Bioactive Compounds Obtained from a Brewery Waste Stream. *Biomed. Res. Int.* **2013**, *2013*, 408491. [CrossRef] [PubMed]
41. Buszewski, B.; Ligor, T.; Ulanowska, A. Determination of volatile organic compounds: Enrichment and analysis. In *Handbook of Trace Analysis*, 1st ed.; Buszewski, B., Ed.; Springer International Publishing: Basel, Switzerland, 2016; pp. 403–430, ISBN 978-3-319-19614-5.
42. Roberto, R.M.; García, N.P.; Hevia, A.G.; Valles, B.S. Application of purge and trap extraction and gas chromatography for determination of minor esters in cider. *J. Chromatogr. A* **2005**, *1069*, 245–251. [CrossRef]
43. Kaczyński, P. Clean-up and matrix effect in LC-MS/MS analysis of food of plant origin for high polar herbicides. *Food Chem.* **2017**, *230*, 524–531. [CrossRef] [PubMed]
44. Dai, J.; Mumper, R.J. Plant Phenolics: Extraction, Analysis and Their Antioxidant and Anticancer Properties. *Molecules* **2010**, *15*, 7313. [CrossRef] [PubMed]
45. Fecka, I.; Turek, S. Determination of polyphenolic compounds in commercial herbal drugs and spices from *Lamiaceae*: Thyme, wild thyme and sweet marjoram by chromatographic techniques. *Food Chem.* **2008**, *108*, 1039–1053. [CrossRef] [PubMed]
46. Žugić, A.; Đorđević, S.; Arsić, I.; Marković, G.; Živković, J.; Jovanović, S.; Tadić, V. Antioxidant activity and phenolic compounds in 10 selected herbs from Vrujci Spa, Serbia. *Ind. Crop. Prod.* **2014**, *52*, 519–527. [CrossRef]
47. Maher, H.M.; Al-Zoman, N.Z.; Al-Shehri, M.M.; Al-Showiman, H.; Al-Taweel, A.M.; Fawzy, G.A.; Perveen, S. Determination of luteolin and apigenin in herbs by capillary electrophoresis with diode array detection. *Instrum. Sci. Technol.* **2015**, *43*, 611–625. [CrossRef]
48. Hawryl, M.A.; Niemiec, M.A.; Słomka, K.; Waksmundzka-Hajnos, M.; Szymczak, G. Two-Dimensional Micro-TLC Phenolic Fingerprints of Selected *Mentha* sp. on Cyano-Bonded Polar Stationary Phase. *J. Chromatogr. Sci.* **2015**, *54*, 64–69. [CrossRef] [PubMed]
49. Arceusz, A.; Wesolowski, M. Quality consistency evaluation of *Melissa officinalis* L. commercial herbs by HPLC fingerprint and quantitation of selected phenolic acids. *J. Pharm. Biomed. Anal.* **2013**, *83*, 215–220. [CrossRef] [PubMed]
50. Skendi, A.; Irakli, M.; Chatzopoulou, P. Analysis of phenolic compounds in Greek plants of *Lamiaceae* family by HPLC. *J. Appl. Res. Med. Aromat. Plants* **2017**, *6*, 62–69. [CrossRef]
51. Vallverdú-Queralt, A.; Regueiro, J.; Alvarenga, J.F.R.; Martinez-Huelamo, M.; Leal, L.N.; Lamuela-Raventos, R.M. Characterization of the phenolic and antioxidant profiles of selected culinary herbs and spices: Caraway, turmeric, dill, marjoram and nutmeg. *Food Sci. Technol.* **2015**, *35*, 189–195. [CrossRef]
52. Taamalli, A.; Arráez-Román, D.; Abaza, L.; Iswaldi, I.; Fernández-Gutiérrez, A.; Zarrouk, M.; Segura-Carretero, A. LC-MS-based metabolite profiling of methanolic extracts from the medicinal and aromatic species *Mentha pulegium* and *Origanum majorana*. *Phytochem. Anal.* **2015**, *26*, 320–330. [CrossRef] [PubMed]
53. Mena, P.; Cirlini, M.; Tassotti, M.; Herrlinger, K.; Dall'Asta, C.; Del Rio, D. Phytochemical Profiling of Flavonoids, Phenolic Acids, Terpenoids, and Volatile Fraction of a Rosemary (*Rosmarinus officinalis* L.) Extract. *Molecules* **2016**, *21*, 1576. [CrossRef] [PubMed]
54. Pandey, R.; Chandra, P.; Kumar, B.; Dutt, B.; Sharma, K.R. A rapid and highly sensitive method for simultaneous determination of bioactive constituents in leaf extracts of six *Ocimum* species using ultra high performance liquid chromatography-hybrid linear ion trap triple quadrupole mass spectrometry. *Anal. Methods* **2016**, *8*, 333–341. [CrossRef]
55. López-Cobo, A.; Gómez-Caravaca, A.M.; Švarc-Gajić, J.; Segura-Carretero, A.; Fernández-Gutiérrez, A. Determination of phenolic compounds and antioxidant activity of a Mediterranean plant: The case of *Satureja montana* subsp. *kitaibelii*. *J. Funct. Foods* **2015**, *18*, 1167–1178. [CrossRef]

56. Cirlini, M.; Mena, P.; Tassotti, M.; Herrlinger, K.A.; Nieman, K.M.; Dall'Asta, C.; Del Rio, D. Phenolic and volatile composition of a dry spearmint (*Mentha spicata* L.) extract. *Molecules* **2016**, *21*, 1007. [CrossRef] [PubMed]
57. Fatiha, B.; Didier, H.; Naima, G.; Khodir, M.; Martin, K.; Léocadie, K.; Caroline, S.; Mohamed, C.; Pierre, D. Phenolic composition, in vitro antioxidant effects and tyrosinase inhibitory activity of three Algerian *Mentha* species: *M. spicata* (L.), *M. pulegium* (L.) and *M. rotundifolia* (L.) Huds (*Lamiaceae*). *Ind. Crop. Prod.* **2015**, *74*, 722–730. [CrossRef]
58. Jesionek, W.; Majer-Dziedzic, B.; Choma, I.M. Separation, Identification, and Investigation of Antioxidant Ability of Plant Extract Components Using TLC, LC-MS, and TLC-DPPH. *J. Liq. Chromatogr. Relat. Technol.* **2015**, *38*, 1147–1153. [CrossRef]
59. Pereira, O.R.; Peres, A.M.; Silva, A.M.; Domingues, M.R.; Cardoso, S.M. Simultaneous characterization and quantification of phenolic compounds in *Thymus x citriodorus* using a validated HPLC-UV and ESI-MS combined method. *Food Res. Int.* **2013**, *54*, 1773–1780. [CrossRef]
60. Hossain, M.B.; Camphuis, G.; Aguiló-Aguayo, I.; Gangopadhyay, N.; Rai, D.K. Antioxidant activity guided separation of major polyphenols of marjoram (*Origanum majorana* L.) using flash chromatography and their identification by liquid chromatography coupled with electrospray ionization tandem mass spectrometry. *J. Sep. Sci.* **2014**, *37*, 3205–3213. [CrossRef] [PubMed]
61. Tuttolomondo, T.; La Bella, S.; Licata, M.; Virga, G.; Leto, C.; Saija, A.; Trombetta, D.; Tomaino, A.; Speciale, A.; Napoli, E.M.; et al. Biomolecular Characterization of Wild Sicilian Oregano: Phytochemical Screening of Essential Oils and Extracts, and Evaluation of Their Antioxidant Activities. *Chem. Biodivers.* **2013**, *10*, 411–433. [CrossRef] [PubMed]
62. Napoli, E.M.; Siracusa, L.; Saija, A.; Speciale, A.; Trombetta, D.; Tuttolomondo, T.; La Bella, S.; Licata, M.; Virga, G.; Leone, R.; et al. Wild Sicilian Rosemary: Phytochemical and Morphological Screening and Antioxidant Activity Evaluation of Extracts and Essential Oils. *Chem. Biodivers.* **2015**, *12*, 1075–1094. [CrossRef] [PubMed]
63. Saija, A.; Speciale, A.; Trombetta, T.; Leto, C.; Tuttolomondo, T.; La Bella, S.; Licata, M.; Virga, G.; Bonsangue, G.; Gennaro, M.C.; et al. Phytochemical, Ecological and Antioxidant Evaluation of Wild Sicilian Thyme: *Thymbra capitata* (L.) Cav. *Chem. Biodivers.* **2016**, *13*, 1641–1655. [CrossRef] [PubMed]
64. Sonmezdag, A.S.; Kelebek, H.; Selli, S. Characterization of aroma-active and phenolic profiles of wild thyme (*Thymus serpyllum*) by GC-MS-Olfactometry and LC-ESI-MS/MS. *J. Food Sci. Technol.* **2016**, *53*, 1957–1965. [CrossRef] [PubMed]
65. Tang, K.S.; Konczak, I.; Zhao, J. Identification and quantification of phenolics in Australian native mint (*Mentha australis* R. Br.). *Food Chem.* **2016**, *192*, 698–705. [CrossRef] [PubMed]
66. Atwi, M.; Weiss, E.K.; Loupassaki, S.; Makris, D.P.; Ioannou, E.; Roussis, V.; Kefalas, P. Major antioxidant polyphenolic phytochemicals of three *Salvia* species endemic to the island of Crete. *J. Herbs Spices Med. Plants* **2016**, *22*, 27–34. [CrossRef]
67. Borrás-Linares, I.; Stojanović, Z.; Quirantes-Piné, R.; Arráez-Román, D.; Švarc-Gajić, J.; Fernández-Gutiérrez, A.; Segura-Carretero, A. *Rosmarinus officinalis* leaves as a natural source of bioactive compounds. *Int. J. Mol. Sci.* **2014**, *15*, 20585–20606. [CrossRef] [PubMed]
68. Çelik, S.E.; Tufan, A.N.; Bekdeşer, B.; Özyürek, M.; Güçlü, K.; Apak, R. Identification and Determination of Phenolics in *Lamiaceae* Species by UPLC-DAD-ESI-MS/MS. *J. Chromatogr. Sci.* **2017**, *55*, 291–300. [CrossRef] [PubMed]
69. Zabot, G.L.; Moraes, M.N.; Rostagno, M.A.; Meireles, M.A.A. Fast analysis of phenolic terpenes by high-performance liquid chromatography using a fused-core column. *Anal. Methods* **2014**, *6*, 7457–7468. [CrossRef]
70. Milevskaya, V.V.; Temerdashev, Z.A.; Butyl'skaya, T.S.; Kiseleva, N.V. Determination of phenolic compounds in medicinal plants from the *Lamiaceae* family. *J. Anal. Chem.* **2017**, *72*, 342–348. [CrossRef]
71. Naczk, M.; Shahidi, F. Extraction and analysis of phenolics in food. *J. Chromatogr. A* **2004**, *1054*, 95–111. [CrossRef]
72. Murkovic, M. Phenolic Compounds: Occurrence, Classes, and Analysis. In *The Encyclopedia of Food and Health*, 1st ed.; Caballero, B., Finglas, P.M., Toldrá, F., Eds.; Academic Press: Kidlington/Oxford, UK, 2016; Volume 4, pp. 346–351, ISBN 978-0-12-384953-3.

73. Oh, Y.; Lee, J.; Yoon, S.; Oh, C.; Choi, D.S.; Choe, E.; Jung, M. Characterization and quantification of anthocyanins in grape juices obtained from the grapes cultivated in Korea by HPLC/DAD, HPLC/MS, and HPLC/MS/MS. *J. Food Sci.* **2008**, *73*, C378–C389. [CrossRef] [PubMed]
74. Kylli, P. Berry Phenolics: Isolation, Analysis, Identification, and Antioxidant Properties. Ph.D. Thesis, University of Helsinki, Helsinki, Finland, 26 August 2011.
75. Bansal, A.; Chhabra, V.; Rawal, R.K.; Sharma, S. Chemometrics: A new scenario in herbal drug standardization. *J. Pharm. Anal.* **2014**, *4*, 223–233. [CrossRef] [PubMed]
76. Areias, F.M.; Valentão, P.; Andrade, P.B.; Moreira, M.M.; Amaral, J.; Seabra, R.M. HPLC/DAD Analysis of Phenolic Compounds from Lavender and its application to quality control. *J. Liq. Chromatogr. Relat. Technol.* **2000**, *23*, 2563–2572. [CrossRef]
77. Proestos, C.; Chorianopoulos, N.; Nychas, G.J.E.; Komaitis, M. RP-HPLC Analysis of the Phenolic Compounds of Plant Extracts. Investigation of Their Antioxidant Capacity and Antimicrobial Activity. *J. Agric. Food Chem.* **2005**, *53*, 1190–1195. [CrossRef] [PubMed]
78. Ćetković, G.S.; Mandić, A.I.; Čanadanović-Brunet, J.M.; Djilas, S.M.; Tumbas, V.T. HPLC Screening of Phenolic Compounds in Winter Savory (*Satureja montana* L.) Extracts. *J. Liq. Chromatogr. Relat. Technol.* **2007**, *30*, 293–306. [CrossRef]
79. Wojdyło, A.; Oszmiański, J.; Czemerys, R. Antioxidant activity and phenolic compounds in 32 selected herbs. *Food Chem.* **2007**, *105*, 940–949. [CrossRef]
80. Lee, J.; Scagel, C.F. Chicoric acid levels in commercial basil (*Ocimum basilicum*) and *Echinacea purpurea* products. *J. Funct. Foods* **2010**, *2*, 77–84. [CrossRef]
81. Kwee, E.M.; Niemeyer, E.D. Variations in phenolic composition and antioxidant properties among 15 basil (*Ocimum basilicum* L.) cultivars. *Food Chem.* **2011**, *128*, 1044–1050. [CrossRef]
82. Šaponjac, V.T.; Čanadanović-Brunet, J.; Ćetković, G.; Djilas, S. Detection of Bioactive Compounds in Plants and Food Products. In *Emerging and Traditional Technologies for Safe, Healthy and Quality Food*, 1st ed.; Nedović, V., Raspor, P., Lević, J., Tumbas Šaponjac, V., Barbosa-Cánovas, G.V., Eds.; Springer International Publishing: Basel, Switzerland, 2016; pp. 81–109, ISBN 978-3-319-24040-4.
83. Yan, R.; Cao, Y.; Yang, B. HPLC-DPPH screening method for evaluation of antioxidant compounds extracted from Semen Oroxyli. *Molecules* **2014**, *19*, 4409–4417. [CrossRef] [PubMed]
84. Srivastava, S.; Adholeya, A.; Conlan, X.A.; Cahill, D.M. Acidic Potassium Permanganate Chemiluminescence for the Determination of Antioxidant Potential in Three Cultivars of *Ocimum basilicum*. *Plant Foods Hum. Nutr.* **2016**, *71*, 72–80. [CrossRef] [PubMed]
85. Fekete, S.; Schappler, J.; Veuthey, J.L.; Guillarme, D. Current and future trends in UHPLC. *TrAC Trends Anal. Chem.* **2014**, *63*, 2–13. [CrossRef]
86. Yuk, J.; Patel, D.N.; Isaac, G.; Smith, K.; Wrona, M.; Olivos, H.J.; Yu, K. Chemical Profiling of Ginseng Species and Ginseng Herbal Products Using UPLC/QTOF-MS. *J. Braz. Chem. Soc.* **2016**, *27*, 1476–1483. [CrossRef]
87. Lu, M.; Yuan, B.; Zeng, M.; Chen, J. Antioxidant capacity and major phenolic compounds of spices commonly consumed in China. *Food Res. Int.* **2011**, *44*, 530–536. [CrossRef]
88. Redruello, B.; Ladero, V.; Del Rio, B.; Fernández, M.; Martin, M.C.; Alvarez, M.A. A UHPLC method for the simultaneous analysis of biogenic amines, amino acids and ammonium ions in beer. *Food Chem.* **2017**, *217*, 117–124. [CrossRef] [PubMed]
89. Wu, N.; Clausen, A.M. Fundamental and practical aspects of ultrahigh pressure liquid chromatography for fast separations. *J. Sep. Sci.* **2007**, *30*, 1167–1182. [CrossRef] [PubMed]
90. Zeng, J.; Zhang, X.; Guo, Z.; Feng, J.; Zeng, J.; Xue, X.; Liang, X. Separation and identification of flavonoids from complex samples using off-line two-dimensional liquid chromatography tandem mass spectrometry. *J. Chromatogr. A* **2012**, *1220*, 50–56. [CrossRef] [PubMed]
91. Kivilompolo, M.; Hyötyläinen, T. Comprehensive two-dimensional liquid chromatography in analysis of *Lamiaceae* herbs: Characterisation and quantification of antioxidant phenolic acids. *J. Chromatogr. A* **2007**, *1145*, 155–164. [CrossRef] [PubMed]
92. Singh, B.; Kumar, A.; Malik, A.K. Flavonoids biosynthesis in plants and its further analysis by capillary electrophoresis. *Electrophoresis* **2016**, *38*, 820–832. [CrossRef] [PubMed]
93. Arries, W.J.; Tredoux, A.G.; Beer, D.; Joubert, E.; Villiers, A. Evaluation of capillary electrophoresis for the analysis of rooibos and honeybush tea phenolics. *Electrophoresis* **2017**, *38*, 897–905. [CrossRef] [PubMed]

94. Patel, K.N.; Patel, J.K.; Patel, M.P.; Rajput, G.C.; Patel, H.A. Introduction to hyphenated techniques and their applications in pharmacy. *Pharm. Methods* **2010**, *1*, 2–13. [CrossRef]
95. Santos, J.; Oliveira, M.B.P.P. Chromatography. In *Food Authentication: Management, Analysis and Regulation*, 1st ed.; Georgiou, C.A., Danazis, G.P., Eds.; John Wiley & Sons Ltd.: Chichester, UK, 2017; pp. 199–232, ISBN 978-1-118-81026-2.
96. Gross, J.H. *Mass Spectrometry: A Textbook*, 2nd ed.; Springer: Berlin/Heidelberg, Germany, 2011; pp. 1–17, ISBN 978-3-642-10711-5.
97. Clarke, W. Mass spectrometry in the clinical laboratory: Determining the need and avoiding pitfalls. In *Mass Spectrometry for the Clinical Laboratory*, 1st ed.; Nair, H., Clarke, W., Eds.; Academic Press: San Diego, CA, USA, 2017; pp. 1–15, ISBN 978-0-12-800871-3.
98. Rubert, J.; Zachariasova, M.; Hajslova, J. Advances in high-resolution mass spectrometry based on metabolomics studies for food—A review. *Food Addit. Contam. Part A* **2015**, *32*, 1685–1708. [CrossRef] [PubMed]
99. Somogyi, Á. Mass spectrometry instrumentation and techniques. In *Medical Applications of Mass Spectrometry*, 1st ed.; Vékey, K., Telekes, A., Vertes, A., Eds.; Elsevier: Amsterdam, The Netherlands, 2008; pp. 93–140, ISBN 978-0-444-51980-1.
100. Boros, B.; Jakabová, S.; Dörnyei, Á.; Horváth, G.; Pluhár, Z.; Kilár, F.; Felinger, A. Determination of polyphenolic compounds by liquid chromatography-mass spectrometry in *Thymus* species. *J. Chromatogr. A* **2010**, *1217*, 7972–7980. [CrossRef] [PubMed]
101. Nagy, T.O.; Solar, S.; Sontag, G.; Koenig, J. Identification of phenolic components in dried spices and influence of irradiation. *Food Chem.* **2011**, *128*, 530–534. [CrossRef] [PubMed]
102. Islam, M.N.; Downey, F.; Ng, C.K.Y. Comparative analysis of bioactive phytochemicals from *Scutellaria baicalensis*, *Scutellaria lateriflora*, *Scutellaria racemosa*, *Scutellaria tomentosa* and *Scutellaria wrightii* by LC-DAD-MS. *Metabolomics* **2011**, *7*, 446–453. [CrossRef]
103. Torras-Claveria, L.; Jauregui, O.; Bastida, J.; Codina, C.; Viladomat, F. Antioxidant Activity and Phenolic Composition of Lavandin (*Lavandula x intermedia* Emeric ex Loiseleur) Waste. *J. Agric. Food Chem.* **2007**, *55*, 8436–8443. [CrossRef] [PubMed]
104. Lee, J.; Scagel, C.F. Chicoric acid found in basil (*Ocimum basilicum* L.) leaves. *Food Chem.* **2009**, *115*, 650–656. [CrossRef]
105. Hossain, M.B.; Rai, D.K.; Brunton, N.P.; Martin-Diana, A.B.; Barry-Ryan, C. Characterization of phenolic composition in *Lamiaceae* spices by LC-ESI-MS/MS. *J. Agric. Food Chem.* **2010**, *58*, 10576–10581. [CrossRef] [PubMed]
106. Chaimbault, P. The Modern Art of Identification of Natural Substances in Whole Plants. In *Recent Advances in Redox Active Plant and Microbial Products: From Basic Chemistry to Widespread Applications in Medicine and Agriculture*, 1st ed.; Jacob, C., Kirsch, G., Slusarenko, A., Winyard, P.G., Burkholz, T., Eds.; Springer: Dordrecht, The Netherlands, 2014; pp. 31–94, ISBN 978-94-017-8953-0.
107. Brice, R.W.; Zhang, X.; Colón, L.A. Fused-core, sub-2 μm packings, and monolithic HPLC columns: A comparative evaluation. *J. Sep. Sci.* **2009**, *32*, 2723–2731. [CrossRef] [PubMed]
108. Simirgiotis, M.; Quispe, C.; Areche, C. Sepúlveda, B. Phenolic Compounds in Chilean Mistletoe (*Quintral, Tristerix tetrandus*) Analyzed by UHPLC–Q/Orbitrap/MS/MS and Its Antioxidant Properties. *Molecules* **2016**, *21*, 245. [CrossRef] [PubMed]
109. Aceña, J.; Stampachiacchiere, S.; Pérez, S.; Barceló, D. Advances in liquid chromatography–high-resolution mass spectrometry for quantitative and qualitative environmental analysis. *Anal. Bioanal. Chem.* **2015**, *407*, 6289–6299. [CrossRef] [PubMed]
110. Kaufmann, A. Combining UHPLC and high-resolution MS: A viable approach for the analysis of complex samples? *TrAC Trends Anal. Chem.* **2014**, *63*, 113–128. [CrossRef]
111. Kennedy, J.; Shanks, K.G.; Van Natta, K.; Conaway, M.C.P.; Wiseman, J.M.; Laughlin, B.; Kozak, M. Rapid screening and identification of novel psychoactive substances using PaperSpray interfaced to high resolution mass spectrometry. *Clin. Mass Spectrom.* **2016**, *1*, 3–10. [CrossRef]
112. Peterman, S.M.; Duczak, N., Jr.; Kalgutkar, A.S.; Lame, M.E.; Soglia, J.R. Application of a Linear Ion Trap/Orbitrap Mass Spectrometer in Metabolite Characterization Studies: Examination of the Human Liver Microsomal Metabolism of the Non-Tricyclic Anti-Depressant Nefazodone Using Data-Dependent Accurate Mass Measurements. *J. Am. Soc. Mass Spectrom.* **2006**, *17*, 363–375. [CrossRef] [PubMed]

113. Li, D.; Schmitz, O.J. Comprehensive two-dimensional liquid chromatography tandem diode array detector (DAD) and accurate mass QTOF-MS for the analysis of flavonoids and iridoid glycosides in *Hedyotis diffusa*. *Anal. Bioanal. Chem.* **2015**, *407*, 231–240. [CrossRef] [PubMed]
114. Senyuva, H.Z.; Gökmen, V.; Sarikaya, E.A. Future perspectives in Orbitrap™-high-resolution mass spectrometry in food analysis: A review. *Food Addit. Contam. Part A* **2015**, *32*, 1568–1606. [CrossRef] [PubMed]

© 2018 by the authors. Licensee MDPI, Basel, Switzerland. This article is an open access article distributed under the terms and conditions of the Creative Commons Attribution (CC BY) license (http://creativecommons.org/licenses/by/4.0/).

Communication

Biological Activities of Extracts from Aerial Parts of *Salvia pachyphylla* Epling Ex Munz

Gabriela Almada-Taylor [1], Laura Díaz-Rubio [1], Ricardo Salazar-Aranda [2], Noemí Waksman de Torres [2], Carla Uranga-Solis [3], José Delgadillo-Rodríguez [4], Marco A. Ramos [1], José M. Padrón [5], Rufina Hernández-Martínez [3],* and Iván Córdova-Guerrero [1],*

[1] Facultad de Ciencias Químicas e Ingeniería, Universidad Autónoma de Baja California, Calzada Universidad 14418, 22390 Tijuana, Mexico; gmaltay@gmail.com (G.A.-T.); ldiaz26@uabc.edu.mx (L.D.-R.); mramos@uabc.edu.mx (M.A.R.)
[2] Departamento de Química Analítica, Facultad de Medicina, Universidad Autónoma de Nuevo León, Madero y Dr. Aguirre Pequeño, 64460 Monterrey, Mexico; salazar121212@yahoo.com.mx (R.S.-A.); nwaksman@gmail.com (N.W.d.T.)
[3] Centro de Investigación Científica y Educación Superior de Ensenada (CICESE), Carretera Tijuana-Ensenada 3918, 22860 Ensenada, Mexico; urangacarla@gmail.com
[4] Facultad de Ciencias, Universidad Autónoma de Baja California, Carretera Tijuana-Ensenada km. 103, 22860 Ensenada, Mexico; jdelga@uabc.edu.mx
[5] Instituto Universitario de Bio-Orgánica "Antonio González", Universidad de La Laguna, Avenida Astrofísico Francisco Sánchez 2, 38206 La Laguna, Spain; jmpadron@ull.edu.es
* Correspondence: ruhernan@cicese.mx (R.H.-M.); icordova@uabc.edu.mx (I.C.-G.); Tel.: +52-664-120-77-41 (I.C.-G.)

Received: 18 October 2018; Accepted: 18 November 2018; Published: 23 November 2018

Abstract: The antioxidant, antimicrobial, antiproliferative, and enzyme inhibitory properties of five extracts from aerial parts of *Salvia pachyphylla* Epling ex Munz were examined to assess the prospective of this plant as a source of natural products with therapeutic potential. These properties were analyzed by performing a set of standard assays. The extract obtained with dichloromethane showed the most variety of components, as they yielded promising results in all completed assays. Furthermore, the extract obtained with ethyl acetate exhibited the greatest antioxidant activity, as well as the best xanthine oxidase inhibitory activity. Remarkably, both extracts obtained with *n*-hexane or dichloromethane revealed significant antimicrobial activity against the Gram-positive bacteria; additionally, they showed greater antiproliferative activity against three representative cell lines of the most common types of cancers in women worldwide, and against a cell line that exemplifies cancers that typically develop drug resistance. Despite that, other extracts were less active, such as the methanolic or aqueous; their results are promising for the isolation and identification of novel bioactive molecules.

Keywords: *Salvia pachyphylla*; plant extracts; antioxidant; antimicrobial; antiproliferative; enzyme inhibitory

1. Introduction

The genus *Salvia* belongs to a large family of flowering plants, Lamiaceae, which comprises about 252 genera and 7200 species [1,2]. Several species of *Salvia* are cultivated for their aromatic features and serve as flavorings, food condiments, cosmetics and perfume additives, and folk medicines [3]. Considering the latter, scrutiny of their chemical constituents have revealed the presence of a vast assortment of active compounds, some of them with antibacterial [4–7], antiviral [8,9], antitumor [10–13], antioxidant [14–17], antidiabetic [18,19], and antiparasitic [20] properties. Additionally, some species have been used etnopharmacologically for the treatment of mental and nervous illness [21], as well as for gastrointestinal conditions [22,23]. Furthermore, phytochemical studies have led to the isolation of many types of diterpenoids, such as abietane, ictexane, labdane, neoclerodane, and phenalenone [24–26], triterpenes and sterols [27], along with anthocyanins, coumarins, polysaccharides, flavonoids, and phenolic acids [22].

Salvia pachyphylla Epling ex Munz (blue sage) is a perennial herbaceous plant distributed from the state of California (USA) to the peninsula of Baja California (Mexico) [28]. The traditional medicine of Native American communities has taken advantage of the curative goods of blue sage and, currently, serves to treat flu symptoms, menstrual depression, and hysteria [29]. Several abietane diterpenoids with pharmacological properties have been isolated from the aerial parts of *S. pachyphylla* [30]. Considering the therapeutic potential of this plant, our study was directed towards identifying specific biological activities existing in different extracts from the aerial parts of *S. pachyphylla*. This approach represents the initial stage of a major survey aimed to isolate and identify phytochemicals with pharmacological potential.

Despite the recent dominance of synthetic chemistry as the foremost method to generate new or improved therapeutic agents, there is still a potential for plants to serve as a natural source of novel drugs [31]. Interestingly, the chemical diversity of natural products is complementary to the diversity found in synthetic libraries. However, natural products are sterically more complex and have a greater diversity because of their long evolutionary selection process [32].

Examples of successful medicines derived from natural products include antibiotics, enzyme inhibitors, immunosuppressive drugs, and antiparasitic agents [33]. The antitumor area is likely the greatest impact of drugs derived from plants, where vinblastine, vincristine, taxol, and camptothecin have improved the effectiveness of chemotherapy of some of the deadliest cancers.

Here, we reported the antioxidant, antimicrobial, antiproliferative, and enzyme inhibition properties of five extracts (obtained with n-hexane, dichloromethane, ethyl acetate, methanol, and water) from the aerial parts of *S. pachyphylla*. These properties were examined performing a set of standard in vitro assays.

2. Results

2.1. Antioxidant Screening

The antioxidant activity was evaluated using the β-carotene-linoleic acid assay and the 1,1-diphenyl-2-picrylhydrazyl DPPH radical-scavenging capacity assay (Figures 1 and 2). In the β-carotene-linoleic test, the best activity was detected in the ethyl acetate extract (84%), immediately followed by the dichloromethane extract (83%); all five extracts showed higher activity, as compared with the reference compound α-tocopherol (8%). On the other hand, in the DPPH system, the ethyl acetate extract remained at the top in this activity, revealing an EC_{50} of 0.28 mg/mL. In addition, the extracts obtained with n-hexane or water showed similar values (0.41 and 0.51 mg/mL, respectively). Remarkably, neither of the extracts exhibited a comparative value with quercetin, the reference compound (0.003 mg/mL).

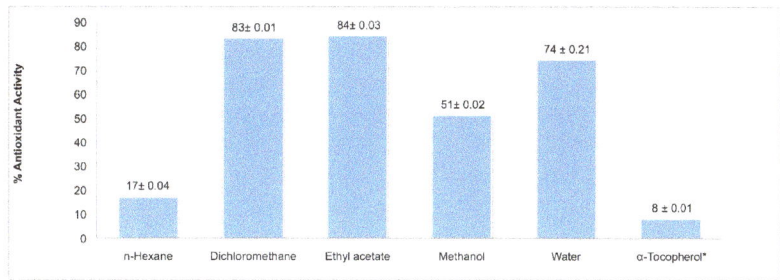

Figure 1. β-carotene-linoleic acid assay of extracts from aerial parts of *Salvia pachyphylla*. * Used as a reference compound. Values are mean ± SD, $n = 3$.

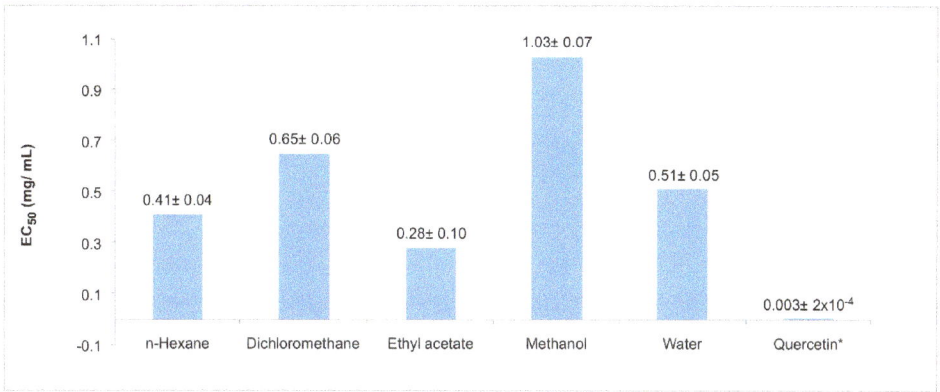

Figure 2. DPPH radical-scavenging capacity assay of extracts from aerial parts of *S. pachyphylla*. * Used as a reference compound. Values are mean ± SD, $n = 3$.

2.2. Antimicrobial Activity

The antimicrobial activity was examined by determining the minimum inhibitory concentrations (MIC) using five bacterial strains and three antibiotics as the reference (Table 1). Interestingly, the extracts obtained with n-hexane or dichloromethane showed significant activity against the Gram-positive *Staphylococcus aureus* and *Enterococcus faecalis*, as well as for the Gram-negative *Escherichia coli*. Furthermore, *E. coli* also exhibited considerable sensitivity to the ethyl acetate extract. Remarkably, the methanolic and the aqueous extracts were inactive against the all bacteria tested. Moreover, *Klebsiella pneumoniae* and *Acinetobacter baumannii* were insensitive to all *S. pachyphylla* extracts examined.

Table 1. In vitro antimicrobial activity of extracts from aerial parts of *S. pachyphylla*.

Antimicrobial Activity (Minimum Inhibitory Concentrations (MIC), μg/mL)						
Extracts or Controls	S. aureus	ORSA [a]	E. faecalis	E. coli	K. pneumoniae	A. baumannii
n-Hexane	62.5	125	250	250	>1000	>1000
Dichloromethane	62.5	125	250	250	>1000	>1000
Ethyl acetate	1000	250	>1000	250	>1000	>1000
Methanol	1000	>1000	>1000	>1000	>1000	>1000
Water	1000	>1000	>1000	>1000	>1000	>1000
Oxacillin *	0.48	125	31.2	0.487	>1000	>1000
Cephalothin *	0.48	62.5	31.2	1	>1000	62.5
Vancomycin *	0.48	1.95	1.95	>250	>1000	250

* Used as a reference compound. [a] Oxacillin-resistant *Staphylococcus aureus*.

2.3. Xanthine and Acetylcholinesterase Inhibitory Assay

The enzymatic evaluation results are shown on Table 2. In the acetylcholinesterase inhibition assay, the extracts did not show a remarkable activity, only the dichloromethane extract presented a slight activity with an IC_{50} of 191.7 µg/mL; however, such a result is far away from the positive control galantamine (0.278 µg/mL). In the xanthine oxidase inhibition assay, better results were obtained, with IC_{50} values for the ethyl acetate and methanol extracts standing out with 11.7 and 19.5 µg/mL, respectively, although they did not surpass the drug allopurinol, used as control (0.842 µg/mL). The rest of the extracts did not show significant activity.

Table 2. Acetylcholinesterase (AChE) and Xanthine Oxidase (XO) inhibitory activity of the extracts from aerial parts of *S. pachyphylla*.

	IC_{50} (µg/mL)	
Extracts or Controls	AChE	XO
n-hexane	>400	254.5 ± 31.7
Dichloromethane	191.7 ± 13.1	86 ± 6.0
Ethylacetate	314.3 ± 43.2	11.7 ± 2.4
Methanol	>400	19.5 ± 0.47
Water	>400	61.8 ± 0.9
Galantamine *	0.278 ± 0.01	ND
Allopurinol *	ND	0.842 ± 0.078

* Used as a reference compound. Values are mean ± SD, n = 3. ND, not determined.

2.4. Antiproliferative Activity

The antiproliferative activity was obtained by measuring the concentration needed to decrease cell propagation by 50% (GI_{50}) using six human cancer cell lines and three well-known anti-cancer drugs (Table 3). All extracts exhibited a degree of effectiveness against all cell lines tested. Specifically, extracts obtained with dichloromethane or n-hexane were the most active against all the evaluated cell lines, showing GI_{50} values between 5.4 and 11 µg/mL. Both extracts showed higher cytotoxicity against cell lines SW1573, T-47D, and WiDr, with concentrations of 6.6, 11, and 8.6 µg/mL and 7.7, 9.9, and 9.9 µg/mL for n-hexane and dichloromethane extracts, respectively; in both cases, the extracts surpassed the positive control etoposide (GI_{50} of 15, 22, and 23 µg/mL against SW1573, T-47D, and WiDr, respectively) and cisplatin (GI_{50} of 15 and 26 µg/mL against T-47D and WiDr, respectively).

Table 3. Antiproliferative activity of extracts from the aerial parts of *S. pachyphylla*.

	GI_{50} (µg/mL)					
Extracts or Controls	A2780	HBL-100	HeLa	SW1573	T-47D	WiDr
n-hexane	6.0	5.9	6.1	6.6	11	8.6
Dichloromethane	5.4	6.7	8.3	7.7	9.9	9.9
Ethylacetate	6.5	18	40	15	38	53
Methanol	34	64	71	70	>100	>100
Water	52	55	77	74	>100	>100
Cisplatin *	ND	1.9	2.0	3.0	15	26
Etoposide *	ND	2.3	3.0	15	22	23
Camptothecin *	ND	ND	0.6	0.25	2.0	1.8

* Used as a reference compound. ND, not determined. Human cancer cell lines: A2780, ovarian carcinoma; HBL-100, breast carcinoma; HeLa, cervix adenocarcinoma; SW1573, lung carcinoma; T-47D, breast ductal carcinoma; and WiDr, colorectal adenocarcinoma.

3. Discussion

Oxidative stress plays a key role in the development of several pathophysiological conditions, such as neurodegenerative and cardiovascular diseases, cancer, and diabetes; natural antioxidants ingested in the daily diet protect the cells against the damage produced by an excess of reactive oxygen species (ROS) [34–37]. There are several techniques for determining the antioxidant capacity; the difference between these methods lies in the assay principle and the experimental conditions [38]. We worked with two of the most used techniques for the antiradical evaluation: DPPH• and ABTS•+ [39], Their high popularity is due to the speed, reproducibility, and simplicity of their procedures. On the other hand, β-carotene gives information regarding the capacity of lipidic peroxidation inhibition [40]. Several studies suggested a good antioxidant potential from *Salvia* species around the world, mainly because of the presence of diterpenes, such as carnosol (1), rosmanol (2), and isorosmanol (3) (Figure 3) [41]. These three compounds are also described in a phytochemical study of *S. pachyphylla* by Guerrero et al. [30].

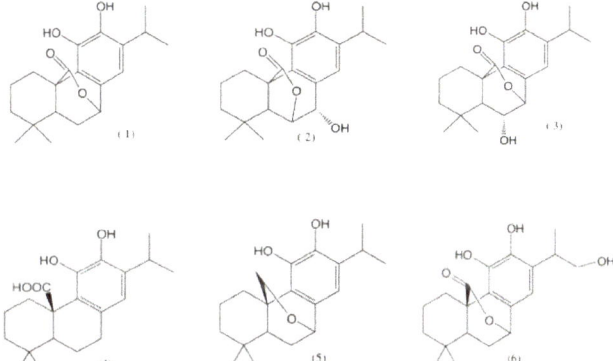

Figure 3. Abietanic diterpenes: carnosol (**1**), rosmanol (**2**), isorosmanol (**3**), carnosic acid (**4**), 20-deoxocarnosol (**5**), and 16-hydroxycarnosol (**6**).

Cuvelier et al. [41] also made an antioxidant evaluation of isolated diterpenes from *Salvia officinalis* such as carnosol and rosmanol by the Method of Antioxidative Power (AOP test) [42], and established that the activity of all these components was related to their phenolic structure. Phenolic diterpenes are widely known to be excellent antioxidants [43,44].

In another study with abietane diterpenes isolated from a dichloromethane extract of *Salvia officinalis* L, made by Miura et al. [45] with the Oil Stability Index (OSI) method [46] and the radical scavenging activity on the DPPH radical, these diterpenes exhibited a remarkably strong antioxidant activity, which was comparable to the one of the standard α-tocopherol. The authors mention that the compounds showing strong antioxidant activity were commonly included an ortho-dihydroxy group in the molecule structure, suggesting that the antioxidant activity is due to the presence of this ortho-dihydroxy group on the C-ring. This characteristic is also found in the metabolites mentioned above that can be found in *S. pachyphylla* extracts.

In the present work, the ethyl acetate extract showed a better antioxidant activity for the two evaluated techniques, along with the dichloromethane extract in the β-carotene assay. These results are in accordance to those obtained by Şenol et al. (2010) and Loizzo et al. (2010) [47,48], where the intermediate polarity extracts (ethyl acetate) from different *Salvia* species presented better results for the same antioxidant techniques of β-carotene and the DPPH radical. The phytochemical background, described above with plants of the *Salvia* genus, suggests that the abietane-type diterpenes can be found in the intermediate polarity extracts of *S. pachyphylla* and could be the responsible for its antioxidant activity.

For the antimicrobial susceptibility analysis, one of the more employed techniques is the calculation of the minimal inhibitory concentration (MIC), which allows researchers to determine easily the minimum quantity of an analyte capable of inhibiting the visible growth of a microorganism [49]. In the antimicrobial evaluation, n-hexane and dichloromethane extracts only matched the MIC of oxacillin in the ORSA (125 µg/mL) and got near to cephalotin with the same bacteria (62.5 µg/mL). Although the reference antibiotics surpassed the extracts activities, it can be noticed that n-hexane and dichloromethane extracts presented a better activity against Gram-positive bacteria *S. aureus* and *E. faecalis*, and only with *E. coli* on the Gram-negative evaluated bacteria. From the higher-polarity extracts (ethyl acetate, methanol, and water), only ethyl acetate presented some degree of activity with ORSA and *E. coli*; however, this was still far away from the reference antibiotics. Methanol and water extracts had even higher MIC than the n-hexane and dichloromethane ones. While our extracts were less active than the references, it can be pointed out that in general, the better results were obtained from the low to medium polarity extracts, against the Gram-positive bacteria. Our results are consistent with those described by Vlietinck et al. [50], who suggested that Gram-positive bacteria are significantly more susceptible to plant-derived extracts. This may be attributed to the fact that the cell wall in Gram-positive bacteria consist of a single layer, while the Gram-negative cell wall is a multilayered and quite complex structure [51]. Previous studies suggested that the antibacterial activity from *Salvia* extracts over Gram-negative bacteria such as *E. coli* depends on the nature of the studied extract [52]. As has been mentioned before, the biological activity of the *Salvia* extracts is related to the presence of abietane diterpenes [53]; in a bioassay-guided study of *Rosmarinus officinalis* L, several extracts were evaluated against different microorganisms responsible for initiating dental caries, *Enterococcus faecalis* among them, and their MICs showed that the higher antibacterial activity was from an extract where carnosic acid (4) and carnosol (1) (Figure 3) were identified as the major compounds by HPLC analysis. As stated before, these metabolites are also found in *S. pachyphylla*.

On the other hand, Oluwatuyi et al. [54], as part of a project to characterize plant-derived natural products that modulate bacterial multidrug resistance (MDR), conducted a bioassay-guided fractionation of a chloroform extract of the aerial parts of *Rosmarinus officinalis*, leading to the characterization of the diterpenes carnosic acid (1) and carnosol (2). The antibacterial activities of these natural products make them interesting potential targets for synthesis.

The screening of natural products in the search of medically relevant enzyme inhibitors remains a viable approach for isolation of novel compounds with specific pharmacological properties, which still has a great potential for further studies. Here, two activity assays were used to identify enzyme inhibitors, within each extract, with therapeutic potential (Table 2): xanthine oxidase (XO) and acetylcholinesterase (AChE) inhibition assays. Nowadays, the potential of natural products has not yet been explored in the search of new treatments for the control of Xanthine Oxidase related diseases [55]. Xanthine oxidase catalyzes the oxidation of hypoxanthine to xanthine and uric acid; however, under certain conditions, this can generate superoxide. It has been proved that XO inhibitors can be helpful for the treatment of liver disease and gout [56]. Acetylcholinesterase catalyzes the hydrolysis of the neurotransmitter acetylcholine into choline and acetic acid [57]. Modulation of acetylcholine levels using acetylcholinesterase inhibitors is among the major strategies to address diverse neurodegenerative diseases [58]. Remarkably, extracts with a prospective inhibitory effect showed a concentration-dependent trend and IC_{50} values were estimated, with the ethyl acetate extract that exhibited a significant effect over XO, while the dichloromethane extract showed a considerable effect on AChE. Unfortunately, other extracts showed a little or poor after-effect on either of the tested enzyme activities, hence, they were considered as inactive.

Regarding the antiproliferative activity, the effects of the n-hexane and dichloromethane extracts over A2780 (ovarian carcinoma), HBL-100 (breast carcinoma), and HeLa (cervix adenocarcinoma), which represent three of the most common cancers in women worldwide, were noticeable. Remarkable results were also obtained against colorectal adenocarcinoma (WiDr), which exemplifies cell lines that typically show drug resistance [59]. Interestingly, these results are different from those previously reported by Guerrero et al. [30]. They tested pure diterpenes isolated from the aerial parts of two species, *S. clevelandi* and *S. pachyphylla*, which were less effective than our extracts. According to the results obtained, from all tested diterpenes, the most active were carnosol (1), 20-deoxocarnosol (5), and 16-hydroxycarnosol (6) (Figure 3) for five different cell lines. We thought that the main difference resides in the nature of the sample, as our results were generated using total extracts, suggesting a possible synergistic effect. Despite other extracts being less active, their results remain promising for pursuing novel molecules with cytotoxic effect.

4. Materials and Methods

4.1. Plant Material

The aerial parts (leaf, flower, and stem) of *S. pachyphylla* were collected in lands of the Sierra Juarez-Constitution National Park, Ensenada, B.C., México, at an elevation of 1630 m and coordinates of N: 32°01′41″; W: 115°56′11″ (Figure 4). Identification and authentication were carried out by Dr. José Delgadillo, and the voucher specimen (BCMEX9783) was deposited in the herbarium of the Autonomous University of Baja California, at Ensenada. Aerial parts (1.3 kg) were air-dried for a week, under shade (to reduce moisture content). The dried material was ground to fine powder and stored at 4 °C until use.

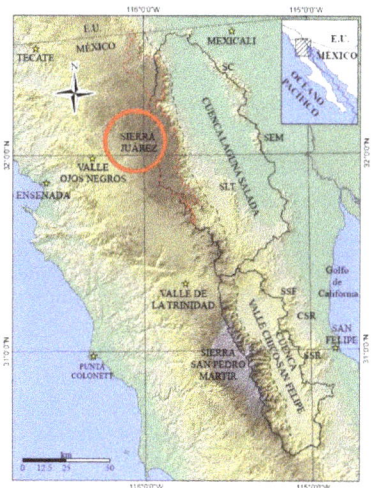

Figure 4. Sierra Juarez-Constitution National Park (marked by red circle), Ensenada, BC, México.

4.2. Preparation Extracts

Crude extracts were obtained through the classical Soxhlet method. Five different thimbles were uniformly packed, each one with 75 g of fine powder. The extraction was carried out using different solvents, one for each one of the thimbles (250 mL): n-hexane (HX), dichloromethane (DC), ethyl acetate (EA), methanol (MT), and distilled water (AQ). The extraction process was stopped until the solvent in siphon tube became colorless. Each extract was filtered and dried at 40 °C, using a rotary evaporator (Buchi Rotavapor® R-215), until a solid or semi-solid residue was yielded. Each residue was separately further lyophilized to get a dry solid matter: HX = 60.3 g, DC = 7.9 g, EA = 24.2 g, MT = 9.9 g, and AQ = 14.5 g. All solids were kept in air tight bottles and stored at 4 °C until use.

4.3. β-Carotene-Linoleic Acid Assay

The antioxidant activity was assayed by the coupled oxidation of β-carotene and linoleic acid as described by Burda and Oleszek [60] with minor modifications. 1 mL of a β-carotene solution

(0.2 mg/mL in chloroform) was added to an emulsion containing 0.018 mL of linoleic acid and 0.2 mL of Tween-20. Chloroform was removed (under a nitrogen environment), 50 mL of aerated deionized water (DO of 9.7 mg/L) was slowly added, and the mixture was vigorously agitated to form a stable emulsion. 5 mL of this emulsion was transferred to test tubes containing the corresponding sample (2 mg) of each extract. Immediately, the absorbance was measured at 470 nm (A_{470}, zero time). All tubes were then incubated at 50 °C and A_{470} values were registered every 15 min for 2 h. A control without the antioxidant was prepared aside and α-tocopherol was used as a reference compound. The antioxidant activity (AA) was expressed as the percentage of inhibition of β-carotene bleaching, as compared to the control, and calculated using the following formula:

$$AA\ (\%) = [1 - (A_S^0 - A_S^{120}/A_C^0 - A_C^{120})] \times 100$$

where A_S and A_C represent the A_{470} value of the sample and control, respectively, and the superscript numbers denote the time of the initial and final measurement (0 and 120 min). All determinations were performed in duplicate and replicated at least three times.

4.4. DPPH Radical-Scavenging Capacity Assay

The radical-scavenging activity was performed as described by Burda and Oleszek [60] with slight modifications. For the evaluation of each extract, a stock solution (4 mg/mL) was prepared and serially two-fold diluted (down to 0.003 mg/mL) with methanol. An aliquot of each dilution (1 mL) was mixed with 1 mL of a methanolic solution of 1,1-diphenyl-2-picrylhydrazyl (DPPH at 0.03 mg/mL). At the same time, a control containing 1 mL of methanol and 1 mL of the DPPH solution was prepared. The mixtures were incubated at room temperature in the dark for 5 min. Using methanol as a blank, the absorbance was quantified at 517 nm (A_{517}). The radical-scavenging activity was calculated as the percentage of DPPH decoloration using the following formula:

$$DPPH\ (\%) = [1 - (A/B) \times 100]$$

where A and B represent the A_{517} value of the control and sample, respectively. All determinations were performed in duplicate and replicated at least three times. For each extract, the percentage of DPPH decoloration was plotted against the concentration of each dilution. The concentration required to decrease the absorbance of DPPH by 50% was obtained by interpolation, from a linear regression analysis, and expresses the EC_{50} value. Quercetin was used as a reference compound.

4.5. Antimicrobial Assay

Antibacterial activity was tested using a microdilution assay following the National Committee for Clinical Laboratory Standards (NCCLS), due its precision and reproducibility [61,62], and MIC was defined as the lowest concentration that prevents visible growth of bacteria. Six strains of microorganisms were used for antimicrobial testing including Gram-positive bacteria (*Enterococcus faecalis*, *Staphylococcus aureus*, and Oxacillin-resistant *Staphylococcus aureus*) and Gram-negative bacteria (*Acinetobacter baumannii*, *Escherichia coli*, and *Klebsiella pneumoniae*); all strains were provided by the Regional Center for Infection Diseases, School of Medicine, Autonomous University of Nuevo Leon (Monterrey, Mexico). All strains were plated on a Müeller–Hinton agar (Becton Dickinson) and incubated at 37 °C for 24 h. Four or five colony forming units were suspended in saline solution and the optical density was adjusted to the turbidity of the 0.5 McFarland Standard. Working suspensions were prepared by a 1:50 dilution in Müeller–Hinton broth. For the evaluation of each extract, a stock solution (6 mg/mL in 5% DMSO) was prepared and serially two-fold diluted with Müeller–Hinton broth in a 96-well microtiter plate (down to 0.5 µg/mL, final highest DMSO concentration 0.83%). One volume (0.1 mL) of working suspension was added to each well. The antibiotics cephalothin, oxacillin, and vancomycin were used as reference compounds.

Controls without bacterial cells (medium control) and without extract or antibiotic (growth control) were prepared aside. Plates were incubated at 37 °C for 48 h and growth was visually examined.

4.6. Xanthine Oxidase Inhibition Assay

There are several assays employed for the Xanthine oxidase activity quantification, the more widely used is by the spectrophotometric determination of uric acid [63], the final reaction product. In this study, the inhibition of the XO activity was evaluated using the easy and sensitive protocol described by Havlik et al. [64]. A reaction solution containing 0.4 mL of 120 mM phosphate buffer (pH 7.8) and 0.33 mL of 150 mM of xanthine was supplemented with 0.25 mL of inhibitor solution (extract or reference) and mixed thoroughly. The reaction was started by adding 0.02 mL of XO enzyme solution (0.5 U/mL). After 3 min of incubation at 24 °C, the uric acid formation was determined by measuring the absorbance at 295 nm (A_{295}). A reaction without inhibitor was used as control and allopurinol served as a reference compound. The inhibition percentage of XO activity was calculated using the following formula:

$$XO\ inhibition\ (\%) = [1 - (As/Ac)] \times 100$$

where A_S and A_C represent the initial velocity of reactions with sample and control, respectively. All determinations were performed in duplicate and replicated at least three times. The concentration required to decrease the activity of XO by 50% was obtained by interpolation, from a linear regression analysis, and expresses the IC_{50} value.

4.7. Acetylcholinesterase (AChE) Inhibition Assay

The acetylcholinesterase inhibition activity was determined with the technique described by Adewusi et al. [65], which is a modification of the widely known and employed Ellman method [66]. For each determination, wells of a microtiter plate were filled with 25 µL of 15 mM acetylthiocholine iodide (in water), 125 µL of 3 mM DTNB in buffer C (50 mM Tris-HCl, pH 8.0, containing 0.1 M NaCl and 0.02 M $MgCl_2 \cdot 6\ H_2O$), 72.5 µL of buffer B (50 mM Tris-HCl, pH 8.0, containing 0.1% BSA), and 2.5 µL of inhibitor solution (extract or reference, in DMSO) and were mixed thoroughly. Absorbance was measured at 412 nm (A_{412}) every 45 s, three times consecutively. Thereafter, 25 µL of AChE enzyme solution (0.2 U/mL) was added to each well and A_{412} was measured five times consecutively every 45 s. A reaction without inhibitor was used as control and galantamine served as the reference compound. Any increase in absorbance due to the spontaneous hydrolysis of the substrate was corrected by subtracting the A_{412} before adding the enzyme. The inhibition percentage of AChE activity was calculated using the following formula:

$$AChE\ inhibition\ (\%) = [1 - (As/Ac)] \times 100$$

where A_S and A_C represent the initial velocity of reactions with sample and control, respectively. All determinations were performed in duplicate and replicated at least three times. The concentration required to decrease the activity of AChE by 50% was obtained by interpolation, from a sigmoidal regression analysis, and expresses the IC_{50} value.

4.8. Cell lines and Culture Conditions

Five human cancer cell lines were used in this study: A2780 (ovarian carcinoma), HBL-100 (breast carcinoma), HeLa (cervix adenocarcinoma), SW1573 (lung carcinoma), T-47D (breast ductal carcinoma), and WiDr (colorectal adenocarcinoma). All line cells were maintained in RPMI (Roswell Park Memorial Institute) 1640 media supplemented with 5% heat-inactivated FCS (Fetal Calf Serum) and 2 mM L-glutamine at 37 °C, 5% CO2, and 95% humidity. Exponentially growing cells were trypsinized and resuspended in medium containing 2% FCS and antibiotics (100 U/mL of penicillin

G and 0.1 mg/mL of streptomycin). Single cell suspensions showing >97% viability, by trypan blue dye exclusion assay, were subsequently counted. After counting, dilutions were made to give the appropriate cell densities required for antiproliferative testing.

4.9. Antiproliferative Assay

Antiproliferative testing was performed using the Sulforhodamine B assay (SRB), the preferred high-throughput assay from National Cancer Institute (NCI, NIH, USA), as reported by Miranda et al. [59], with slight modifications. The SRB assay is based on the dye union to the basic amino acids of the cellular proteins, and the colorimetric evaluation provides an estimation of the total protein mass, which is related to the number of cells. This assay presents an excellent linearity, sensibility, and low costs compared to others [67]. Each extract was initially dissolved in DMSO at 400 times the desired maximum concentration to test. Six thousand cells were inoculated to each well of a microtiter plate (100 µL of a suspension of 6×10^4 cells per mL). One day after plating, all testing samples and reference compounds were added to corresponding wells (triplicated). Samples were tested in six decimal serial dilutions starting at 250 µg/mL. Control cultures were tested against equivalent concentrations of DMSO (0.25% as a negative control). After 48 h of incubation at permissive conditions, cell cultures were treated with 25 µL ice-cold 50% TCA and fixed at 4 °C for 60 min. Thereafter, the SRB assay was performed. The absorbance of each well was measured at 492 nm (A_{492}). All absorbance values were corrected for background A_{492} (control wells containing just culture media). As recommended by the NCI, the concentration that causes 50% growth inhibition, GI_{50} value, was corrected by count at time zero; thus, GI_{50} is the concentration where $[(T - T_0)/(C - T_0)] = 0.5$. The absorbance of the test well after 48 h is T, the absorbance at time zero is T_0, and the absorbance of the control is C. For all these calculations, Excel spreadsheets were used. GI_{50} values were computed from the dose-response curves.

5. Conclusions

In conclusion, aerial parts of *S. pachyphylla* have prospective potential as a source of natural products that could act as antioxidants, antimicrobial and antiproliferative compounds, or enzyme inhibitors.

Although our findings could be the outcome of a synergistic effect, they support the notion of aiming our next approach towards the isolation and identification of novel molecules with therapeutic potential.

Author Contributions: G.A.-T., performed the investigation and wrote the original draft of the paper; L.D.-R., R.S.-A., J.D.-R., and J.M.P. adjusted the methodology of the different experiments realized.; N.W.d.T., J.M.P., and I.C.-G. did the formal analysis of the results generated.; M.A.R., R.H.-M., and C.U.-S. did the review and editing of the paper.; R.H.-M. managed the funding acquisition.; I.C.-G. conducted the supervision and project administration.

Funding: This research received no external funding.

Acknowledgments: The authors would like to thank the Facultad de Ciencias Químicas e Ingeniería, Universidad Autónoma de Baja California, for the support given and for allowing the use of their facilities during this investigation. Additionally, to CONACYT, who supported G.A.-T. with a scholarship during her PhD studies during the realization of this investigation.

Conflicts of Interest: The authors declare no conflict of interest.

References

1. Mabberley, D.J. *The Plant-Book, a Portable Dictionary of Vascular Plants*, 2nd ed.; University of Cambridge: Cambridge, UK, 1997; pp. 385–635. ISBN 0521414210.
2. Harley, R.M.; Atkins, S.; Budantsev, A.L.; Cantino, P.D.; Conn, B.J.; Grayer, R.; Harley, M.M.; De Kok, R.P.J.; Krestovskaja, T.; Morales, R. Labiatae. In *The Families and Genera of Vascular Plants*, 1st ed.; Springer: Berlin, Germany, 2004; pp. 167–282. ISBN 3540405933.
3. Firdous, S.; Dadass, A.K.; Khan, K.M.; Usmani, S.B.; Ahmad, V.U. A new triterpenoid from the leaves of *Salvia triloba*. *Fitoterapia* **1999**, *70*, 326–327. [CrossRef]

4. Ikram, M.; Haq, I. Screening of medicinal plants for antimicrobial activity. *Fitoterapia* **1980**, *51*, 231–235.
5. González, A.G.; Abad, T.; Jimenez, I.A.; Ravelo, A.G.; Zahira Aguiar, J.G.L.; San Andres, L.; Plasencia, M.; Herrera, J.R.; Moujir, L. A first study of antibacterial activity of diterpenes isolated from some *Salvia* species (Lamiaceae). *Biochem. Syst. Ecol.* **1989**, *7*, 293–296. [CrossRef]
6. Moujir, L.; Gutierrez-Navarro, A.M.; San Andres, L.; Luis, J.G. Structure-antimicrobial activity relationships of abietane diterpenes from *Salvia* species. *Phytochemistry* **1993**, *34*, 1493–1495. [CrossRef]
7. Moujir, L.; Gutierrez-Navarro, A.M.; San Andres, L.; Luis, J.G. Bioactive diterpenoids isolated from *Salvia mellifera*. *Phytother. Res.* **1996**, *10*, 172–174. [CrossRef]
8. Cui, X.Y.; Wang, Y.L.; Kokudo, N.; Fang, D.Z.; Tang, W. Traditional Chinese medicine and related active compounds against hepatitis B virus infection. *Biosci. Trends* **2010**, *4*, 39–47. [PubMed]
9. Alim, A.; Goze, I.; Goze, H.M.; Tepe, B. In vitro antimicrobial and antiviral activities of the essential oil and various extracts of *Salvia cedronella* Boiss. *J. Med. Plants Res.* **2009**, *3*, 413–419.
10. Hu, S.; Chen, S.M.; Li, X.K.; Qin, R.; Mei, Z.N. Antitumor effects of Chi-Shen extract from *Salvia miltiorrhiza* and Paeoniae radix on human hepatocellular carcinoma cells. *Acta Pharmacol. Sin.* **2007**, *28*, 1215–1223. [CrossRef] [PubMed]
11. Dat, N.T.; Jin, X.; Lee, J.H.; Lee, D.; Hong, Y.S.; Lee, K.; Kim, Y.H.; Lee, J.J. Abietane diterpenes from *Salvia miltiorrhiza* inhibit the activation of hypoxia-inducible factor-1. *J. Nat. Prod.* **2007**, *70*, 1093–1097. [CrossRef] [PubMed]
12. Wang, X.H.; Bastow, K.F.; Sun, C.M.; Lin, Y.L.; Yu, H.J.; Don, M.J.; Wu, T.S.; Nakamura, S.; Lee, K.H. Antitumor agents. 239. Isolation, structure elucidation, total synthesis, and anti-breast cancer activity of neo-tanshinlactone from *Salvia miltiorrhiza*. *J. Med. Chem.* **2004**, *47*, 5816–5819. [CrossRef] [PubMed]
13. Ulubelen, A.; Topcu, G.; Tan, N.; Lin, L.J.; Cordell, G.A. Microstegiol, a rearranged diterpene from *Salvia microstegia*. *Phytochemistry* **1992**, *31*, 2419–2421. [CrossRef]
14. Kabouche, A.; Kabouche, Z.; Öztürk, M.; Kolak, U.; Topçu, G. Antioxidant abieetane diterpenoids from *Salvia barrelieri*. *Food Chem.* **2007**, *102*, 1281–1287. [CrossRef]
15. Weng, X.C.; Wang, W. Antioxidant activity of compounds isolated from *Salvia plebeia*. *Food Chem.* **2000**, *71*, 489–493. [CrossRef]
16. Tepe, B.; Sokmen, M.; Akpulat, H.A.; Sokmen, A. Screening of the antioxidant potentials of six *Salvia* species from Turkey. *Food Chem.* **2006**, *95*, 200–204. [CrossRef]
17. Kamatou, G.P.P.; Viljoen, A.M.; Steenkamp, P. Antioxidant, antiinflammatory activities and HPLC analysis of South African *Salvia* species. *Food Chem.* **2010**, *119*, 684–688. [CrossRef]
18. Hitokoto, H.; Morozumi, S.; Wauke, T.; Saiki, S.; Kurata, H. Inhibitory effects of spices on growth and toxin production of toxigenic fungi. *Appl. Environ. Microbiol.* **1980**, *39*, 818–822. [PubMed]
19. Han, Y.M.; Oh, H.; Na, M.; Kim, B.S.; Oh, W.K.; Kim, B.Y.; Jeong, D.G.; Ryu, S.E.; Sok, D.E.; Ahn, J.S. PTP1B inhibitory effect of abietane diterpenes isolated from *Salvia miltiorrhiza*. *Biol. Pharm. Bull.* **2005**, *28*, 1795–1797. [CrossRef] [PubMed]
20. Sairafianpour, M.; Christensen, J.; Staerk, D.; Budnik, B.A.; Kharazmi, A.; Bagherzadeh, K.; Jaroszewski, J.W. Leishmanicidal, antiplasmodial, and cytotoxic activity of novel diterpenoid 1,2-quinones from *Perovskia abrotanoides*: New source of tanshinones. *J. Nat. Prod.* **2001**, *64*, 1398–1403. [CrossRef] [PubMed]
21. Maklad, Y.A.; Aboutabl, E.A.; El-Sherei, M.M.; Meselhy, K.M. Bioactivity studies of *Salvia transsylvanica* (Schur ex Griseb) grown in Egypt. *Phytother. Res.* **1999**, *13*, 147–150. [CrossRef]
22. Lu, Y.; Foo, L.Y. Polyphenolics of *Salvia*. A review. *Phytochemistry* **2002**, *59*, 117–140. [CrossRef]
23. Tepe, B.; Dönmez, E.; Unlu, M.; Candan, F.; Dimitra, D.; Vardar-Unlu, G. Antimicrobial and antioxidative activities of the essential oils and methanol extracts of *Salvia cryptantha* (Montbret et Aucher ex Benth.) and *Salvia multicaulis* (Vahl). *Food Chem.* **2004**, *84*, 519–525. [CrossRef]
24. Kusumi, T.; Ooi, T.; Hayashi, T.; Kakisawa, H. A diterpenoid phenalenone from *Salvia miltiorrhiza*. *Phytochemistry* **1985**, *24*, 2118–2120. [CrossRef]
25. Habibi, Z.; Eftekhar, F.; Samiee, K.; Rustaiyan, A. Structure and antibacterial activity of a new labdane diterpenoid from *Salvia leriaefolia*. *J. Nat. Prod.* **2000**, *63*, 270–271. [CrossRef] [PubMed]
26. Nieto, M.; García, E.E.; Giordano, O.S.; Tonn, C.E. Icetexane and abietane diterpenoids from *Salvia gilliessi*. *Phytochemistry* **2000**, *53*, 911–915. [CrossRef]

27. Rauter, A.P.; Branco, I.; Lopes, R.G.; Justino, J.; Silva, F.V.M.; Noronha, J.P.; Cabrita, E.J.; Brouard, I.; Bermejo, J. A new lupine triterpenetriol and anticholinesterase activity of *Salvia sclareoides*. *Fitoterapia* **2007**, *78*, 474–481. [CrossRef] [PubMed]
28. Taylor, R. The Origin and Adaptive Radiation of *Salvia pachyphylla* (*Lamiaceae*). Master's Thesis, Northern Arizona University, Flagstaff, Arizona, 2002.
29. Alonso-Castro, A.J.; Villarreal, M.L.; Salazar-Olivo, L.A.; Gomez-Sanchez, M.; Dominguez, F.; Garcia-Carranca, A. Mexican medicinal plants used for cancer treatment: Pharmacological, phytochemical and ethnobotanical studies. *J. Ethnopharmacol.* **2011**, *133*, 945–972. [CrossRef] [PubMed]
30. Guerrero, I.C.; Andrés, L.S.; León, L.G.; Machín, R.P.; Padrón, J.M.; Luis, J.G.; Delgadillo, J. Abietane diterpenoids from *Salvia pachyphylla* and *S. clevelandii* with cytotoxic activity against human cancer cell lines. *J. Nat. Prod.* **2006**, *69*, 1803–1805. [CrossRef] [PubMed]
31. Kviecinski, M.R.; Felipe, K.B.; Schoenfelder, T.; de Lemos Wiese, L.P.; Rossi, M.H.; Gonçalez, E.; Felicio, J.D.; Filho, D.W.; Pedrosa, R.C. Study of the antitumor potential of *Bidens pilosa* (Asteraceae) used in Brazilian folk medicine. *J. Ethnopharmacol.* **2008**, *117*, 69–75. [CrossRef] [PubMed]
32. Rates, S.K.M. Plants as source of drugs. *Toxicon* **2001**, *39*, 603–613. [CrossRef]
33. Harvey, A.L.; Waterman, P. The continuing contribution of biodiversity to drug discovery. *Curr. Opin. Drug Discov. Dev.* **1998**, *1*, 71–76.
34. Ighodaro, O.M. Molecular pathways associated with oxidative stress in diabetes mellitus. *Biomed. Pharmacother.* **2018**, *108*, 656–662. [CrossRef] [PubMed]
35. Klaunig, J.E.; Wang, Z. Oxidative stress in carcinogenesis. *Curr. Opin. Toxicol.* **2018**, *7*, 116–121. [CrossRef]
36. Taleb, A.; Ahmad, K.A.; Ihsan, A.U.; Qu, J.; Lin, N.; Hezam, K.; Koju, N.; Hui, L.; Qilong, D. Antioxidant effects and mechanism of silymarin in oxidative stress induced cardiovascular diseases. *Biomed. Pharmacother.* **2018**, *102*, 689–698. [CrossRef] [PubMed]
37. Tramutola, A.; Lanzillotta, C.; Perluigi, M.; Butterfield, D.A. Oxidative stress, protein modification and Alzheimer disease. *Brain Res. Bull.* **2017**, *133*, 88–96. [CrossRef] [PubMed]
38. Joon, K.M.; Takayuki, S. Antioxidant Assays for Plant and Food Components. *J. Agric. Food Chem.* **2009**, *57*, 1655–1666. [CrossRef]
39. Reşat, A.; Mustafa, Ö.; Kubilay, G.; Esra, Ç. Antioxidant Activity/Capacity Measurement. 1. Classification, Physicochemical Principles, Mechanisms, and Electron Transfer (ET)-Based Assays. *J. Agric. Food Chem.* **2016**, *64*, 997–1027. [CrossRef]
40. Floegel, A.; Dae-Ok, K.; San-Jin, C.; Sung, I.K.; Ock, K.C. Comparison of ABTS/DPPH assays to measure antioxidant capacity in popular antioxidant-rich US foods. *J. Food Compos. Anal.* **2011**, *24*, 1043–1048. [CrossRef]
41. Cuvelier, M.E.; Berset, C.; Richard, H. Antioxidant Constituents in Sage (*Salvia officinalis*). *J. Agric. Food Chem.* **1994**, *42*, 665–669. [CrossRef]
42. Cuvelier, M.E.; Berset, C.; Richard, H. Use of a new test for determining comparative antioxidative activity of BHA, BHT, alpha- and gamma-tocopherols and extracts from rosemary and sage. *Sci. Aliments* **1990**, *10*, 797–806.
43. Nakatani, N.; Inatani, R. Structure of rosmanol, a new antioxidant from rosemary. *Agric. Biol. Chem.* **1981**, *45*, 2385–2386. [CrossRef]
44. Nakatani, N.; Inatani, R. A new diterpene lactone, rosmadial, from rosemary. *Agric. Biol. Chem.* **1983**, *47*, 353–358. [CrossRef]
45. Miura, K.; Kikuzaki, H.; Nakatani, N. Antioxidant Activity of Chemical Components from Sage (*Salvia officinalis* L.) and Thyme (*Thymus vulgaris* L.) Measured by the Oil Stability Index Method. *J. Agric. Food Chem.* **2002**, *50*, 1845–1851. [CrossRef] [PubMed]
46. Nakatani, N.; Tachibana, Y.; Kikuzaki, H. Establishment of a model substrate oil for antioxidant activity Assessment by Oil Stability Index Method. *J. Am. Oil Chem. Soc.* **2001**, *78*, 19–23. [CrossRef]
47. Şenol, F.S.; Orhan, I.; Celep, F.; Kahraman, A.; Doğan, M.; Yilmaz, G.; Şener, B. Survey of 55 Turkish Salvia taxa for their acetylcholinesterase inhibitory and antioxidant activities. *Food Chem.* **2010**, *120*, 34–43. [CrossRef]
48. Loizzo, M.R.; Tundis, R.; Conforti, F.; Menichini, F.; Bonesi, M.; Nadjafi, F.; Frega, N.G.; Menichini, F. *Salvia leriifolia* Benth (*Lamiaceae*) extract demonstrates in vitro antioxidant properties and cholinesterase inhibitory activity. *Nutr. Res.* **2010**, *30*, 823–830. [CrossRef] [PubMed]

49. Sánchez-García, E.; Castillo-Hernández, S.L.; García-Palencia, P. Actividad Antimicrobiana. In *Investigación en Plantas de Importancia Médica*, 1st ed.; Rivas-Morales, C., Oranday-Cardenas, M.A., Verde-Star, M.J., Eds.; Publisher: Barcelona, España, 2016; Volume 1, pp. 77–100. ISBN 978-84-944673-7-0.
50. Vlietinck, A.J.; Van, H.L.; Totte, J.; Lasure, A.; Vanden, B.D.; Rwangabo, P.C.; Mvukiyumwami, J. Screening of hundred Rwandese medicinal plants for antimicrobial and antiviral properties. *J. Ethnopharmacol.* **1995**, *46*, 31–47. [CrossRef]
51. Abu-Darwish, M.S.; Al-Ramamneh, E.A.D.M.; Kyslychenko, V.S.; Karpiuk, U.V. The antimicrobial activity of essential oils and extracts of some medicinal plants grown in Ash-shoubak region—South of Jordan. *Pak. J. Pharm. Sci.* **2012**, *25*, 239–246. [PubMed]
52. Ghorbani, A.; Esmaeilizadeh, M. Pharmacological properties of *Salvia officinalis* and its components. *J. Tradit. Complement. Med.* **2017**, *7*, 433–440. [CrossRef] [PubMed]
53. González, C.M.A. Aromatic abietane diterpenoids: Their biological activity and synthesis. *Nat. Prod. Rep.* **2015**, *32*, 684–704. [CrossRef] [PubMed]
54. Oluwatuyi, M.; Kaatz, G.W.; Gibbons, S. Antibacterial and resistance modifying activity of *Rosmarinus officinalis*. *Phytochemistry* **2004**, *65*, 3249–5324. [CrossRef] [PubMed]
55. Hudaib, M.M.; Tawaha, K.A.; Mohammad, M.K.; Assaf, A.M.; Issa, A.Y.; Alali, F.Q.; Aburjai, T.A.; Bustanji, Y.K. Xanthine oxidase inhibitory activity of the methanolic extracts of selected Jordanian medicinal plants. *Pharmacogn. Mag.* **2011**, *7*, 320–324. [CrossRef] [PubMed]
56. Sahgal, G.; Ramanathan, S.; Sasidharan, S.; Mordi, M.N.; Ismail, S.; Mansor, S.M. In vitro antioxidant and xanthine oxidase inhibitory activities of methanolic *Swietenia mahagoni* seed extracts. *Molecules* **2009**, *14*, 4476–4485. [CrossRef] [PubMed]
57. Čolović, M.B.; Krstić, D.Z.; Lazarević-Pašti, T.D.; Bondžić, A.M.; Vasić, V.M. Acetylcholinesterase Inhibitors: Pharmacology and Toxicology. *Curr. Neuropharmacol.* **2013**, *11*, 315–335. [CrossRef] [PubMed]
58. Hasnat, M.A.; Pervin, M.; Lim, B.O. Acetylcholinesterase inhibition and in vitro and in vivo antioxidant activities of *Ganoderma lucidum* grown on germinated brown rice. *Molecules* **2013**, *18*, 6663–6678. [CrossRef] [PubMed]
59. Miranda, P.O.; Padrón, J.M.; Padrón, J.I.; Villar, J.; Martín, V.S. Prins-type synthesis and SAR study of cytotoxic alkyl chloro dihydropyrans. *Chem. Med. Chem.* **2006**, *1*, 323–329. [CrossRef] [PubMed]
60. Burda, S.; Oleszek, W. Antioxidant and antiradical activities of flavonoids. *J. Agric. Food Chem.* **2001**, *49*, 2774–2779. [CrossRef] [PubMed]
61. Alanís-Garza, B.A.; González-González, G.M.; Salazar-Aranda, R.; Waksman de Torres, N.; Rivas-Galindo, V.M. Screening of antifungal activity of plants from the northeast of Mexico. *J. Ethnopharmacol.* **2007**, *114*, 468–471. [CrossRef] [PubMed]
62. Salazar-Aranda, R.; Pozos, M.E.; Cordero, P.; Perez, J.; Salinas, M.C.; Waksman, N. Determination of the antioxidant activity of plants from northeast Mexico. *Pharm. Biol.* **2008**, *46*, 166–170. [CrossRef]
63. Shintani, H. Determination of Xanthine Oxidase. *Pharm. Anal. Acta* **2013**, *S7*, 4. [CrossRef]
64. Havlik, J.; de la Huebra, R.G.; Hejtmankova, K.; Fernandez, J.; Simonova, J.; Melich, M.; Rada, V. Xanthine oxidase inhibitory properties of Czech medicinal plants. *J. Ethnopharmacol.* **2010**, *132*, 461–465. [CrossRef] [PubMed]
65. Adewusi, E.A.; Moodley, N.; Steenkamp, V. Antioxidant and acetylcholinesterase inhibitory activity of selected southern African medicinal plants. *S. Afr. J. Bot.* **2011**, *77*, 638–644. [CrossRef]
66. Mathew, M.; Subramanian, S. In Vitro Screening for Anti-Cholinesterase and Antioxidant Activity of Methanolic Extracts of Ayurvedic Medicinal Plants Used for Cognitive Disorders. *PLoS ONE* **2014**, *9*, e86804. [CrossRef] [PubMed]
67. Kuete, V.; Karaosmanoglu, O.; Sivas, H. *Anticancer Activities of African Medicinal Spices and Vegetables In Medicinal Spices and Vegetables from Africa*, 1st ed.; Kuete, V., Ed.; Academic Press: Cambridge, MA, USA, 2017; Volume 1, pp. 271–297. ISBN 9780128092866.

© 2018 by the authors. Licensee MDPI, Basel, Switzerland. This article is an open access article distributed under the terms and conditions of the Creative Commons Attribution (CC BY) license (http://creativecommons.org/licenses/by/4.0/).

Article

Biological Activity of the *Salvia officinalis* L. (Lamiaceae) Essential Oil on *Varroa destructor* Infested Honeybees

Leila Bendifallah [1,*], Rachida Belguendouz [2], Latifa Hamoudi [3] and Karim Arab [4]

1. Laboratory of Soft technology, Valorization, Physical-chemistry of biological materials and Biodiversity, Department of Agronomy, Faculty of Sciences, Université M'hamed Bougara, Boumerdes, Avenue de l'indépendance, Boumerdes 35000, Algeria
2. Laboratory of Aromatic and Medicinal Plants, Biotechnology Department, Faculty of Nature Sciences and Life, University of Blida, Blida 09000, Algeria; belguendouzr@yahoo.com
3. Laboratoire de Technologie Alimentaire, Faculté des Sciences de l'ingénieur, Université M'Hamed Bougara, Avenue de L'indépendance, Boumerdès 35000, Algérie; Latifa.Hamoudi.1@ulaval.ca
4. Laboratoire Valorisation et Conservation des Ressources Biologiques, Department of Agronomy, Faculty of Sciences, Université M'hamed Bougara, Boumerdes 35000, Algeria; arabkarim3@gmail.com
* Correspondence: leila.bendifallah@gmail.com; Tel.: +213-555-285-705

Received: 1 May 2018; Accepted: 5 June 2018; Published: 6 June 2018

Abstract: The present work is conducted as part of the development and the valorization of bioactive natural substances from Algerian medicinal and aromatic spontaneous plants, a clean alternative method in biological control. For this purpose, the bio-acaricidal activity of *Salvia officinalis* (sage)essential oil (EO)was evaluated against the *Varroa destructor*, a major threat to the honey bee *Apis mellifera* ssp. *intermissa*. The aerial parts of *S. officinalis* L., 1753 were collected from the Chrea mountainous area in Northern Algeria. They were subjected to hydro distillation by a Clevenger apparatus type to obtain the EO, and screened for bio-acaricidal activity against *Varroa destructor* by the method of strips impregnated with the mixture EO and twin according to three doses. Pre-treatment results revealed infestation rates in the experimental site ranging from 3.76% to 21.22%. This showed the heterogeneity of infestations in hives according to the density of bees. This constituted a difficulty in monitoring the population dynamics of this parasite. After treatment, a difference in the acaricidal effect of Sage essential oil is noticed. It gives a mortality rate of 6.09% by the dose D1: 5%, 2.32% by the dose D2: 15%, and a low mortality rate of 0.9% by the dose D3: 20%. The chemical treatment carried out by Bayvarol gives a result close to that of the essential oil of Sage (9.97%).These results point to the fact that Sage essential oil treatments have a significant effect and good biological activity with regard to harmful species.

Keywords: *Apis mellifera intermissa*; bio-acaricide; Algeria

1. Introduction

In addition to its different products (honey, royal jelly, pollen), the honeybee plays an important role in the pollination of both cultivated and wild plant species. As a result, it actively contributes to the development and preservation of the biodiversity of ecosystems by promoting the sustainability of flowering plants. It is also considered a true bio-indicator of environmental health [1].

However, the bee is confronted with several constraints that limit its development and that can even cause its disappearance. These constraints are related to the environment (climate and the reduction of food resources), chemical agents (intoxication by phytosanitary products), apicultural practices, and biological agents (bacteria, viruses, parasites, predators).

Among the bee diseases, Varroasisis considered one of the most widespread and dangerous diseases. It is caused by the *Varroa destrucor* Anderson and Trueman, 2000, mite, which parasitizes both brood and adult bees, thus causing considerable losses to the bee population. Several factors, such as the sale of queens and swarms and transhumance, have contributed to the spread of this disease worldwide [2,3].

The losses amount to some thousands of hives just for Eastern Europe [4]. In Algeria, the first varroa was reported in June 1981 on *A. mellifera* ssp. *Intermissa* at Umm-Teboul in Alkala (Annaba). Since then, this parasite has spread to the north of the country [5] and caused the considerable loss of hives. Several methods of physical (heat and fumigation), genetic (breed selection), chemical, and biological control show their efficiency on this parasite, including products like Bayvarol, Apiston, and Apivar and plant extracts such as essential oils of *Eucalyptus Radiata* A. Cunn, *Allium sativum* L., *Rosmarinus officinalis* L., and *Origanum glandulosum* Desf. [6].

Currently, there is lot of interest towards traditional medicines and herbal-based treatment all over the world. Therefore, numerous experimental and clinical studies are being undertaken on medicinal plants and there is a need for updating and integrating the findings [7].

Colin [8] has shown that many plant essential oils have an antiparasite effect, act on the behavior and/or the development of certain arthropods, and can sometimes be fatal.

More than 120 components are characterized in the essential oil prepared from aerial parts of *S. officinalis*. The main components of the oil include borneol, camphor, caryophyllene, cineole, elemene, humulene, ledene, pinene, and thujone [9–11]. The sage oil has a direct effect on the nervous system [12]. Camphor, thujone, and terpene ketones are considered as the most toxic compounds in *S. officinalis*. These compounds may induce toxic effects. According to Veličković et al. [13], bornyl acetate, camphene, camphor, humulene, limonene, and thujone are the most comment phytochemicals in the leaves. However, it should be considered that, like other herbs, the chemical composition of *S. officinalis* varies depending on the environmental conditions such as climate, water availability, and altitude [14].

In this work, *Salvia officinalis* (Lamiaceae) is chosen for its acaricidal effect on the *Varroa destructor*, parasite of *Apis mellifera* ssp. *Intermissa*, in comparison with Apivar, which is a commercial chemical product frequently used in Algeria. Our goal is to study the acaricidal activity of its essential oil as a biological alternative to the biopesticide usually practiced in Algeria.

2. Results

2.1. Estimation of the Initial Infestation Rate in Hives by the Varroa Mite Destructor before Treatment

The estimated global number of bees on which we conducted our study is 85,000 bees (Table 1). The estimated overall parasitic varroa number is 9801, with an infestation rate of 11.53% judged to be higher than the tolerance threshold established by Robaux [15], which is 5%. This is a strong indicator that this apiary deserves support and immediate treatment.

Table 1. Estimation of the initial infestation rate (before treatment) in the different hives and lots.

Hive Number	Number of VARROA DIED for 01 Months "A"	Mean Mortality/d $B = A/29$	Estimated Varroa Population $C = B \times 90$	Estimated Population Bees: P	Initial Infestation Rate T_1%
Hive1	422	14.55	1309	6250	20.94
Hive 2	114	3.93	354	8750	4.04
Witness batch1	536	18.48	1663	15,000	11.087
Hive 3	91	3.13	282	7500	3.76
Hive 4	131	4.51	406	8750	4.64
Batch2	222	7.64	688	16,250	4.23
Hive 5	229	7.87	708	8125	8.71
Hive 6	209	7.2	648	11,250	5.76
Batch3	438	15.07	1356	19,375	6.99

Table 1. Cont.

Hive Number	Number of VARROA DIED for 01 Months "A"	Mean Mortality/d B = A/29	Estimated Varroa Population C = B × 90	Estimated Population Bees: P	Initial Infestation Rate T_1%
Hive 7	385	13.27	1194	5625	21.22
Hive 8	433	14.93	1344	10,000	13.44
Batch4	818	28.2	2538	15,625	16.24
Hives treated per essential oil	2014	69.39	6245	66,250	9.43
Hive 9	644	22.2	1998	10,625	18.8
Hive 10	502	17.31	1558	8125	19.18
Batch5	1146	39.51	3556	18,750	18.96
Batch chemically treated	1146	39.51	3556	18,750	18.97
global apiary (10 hives)	3160	108.9	9801	85,000	11.53

2.2. Estimation of the Infestation Rate in the Different Hives after Treatment with the Essential Oil of Sage Salvia officinalis

We note that the rate of varroa infection estimated after treatment with the essential oil of sage is 6930 individuals living at the expense of 66,250 bees (Table 2). This number is the equivalent of an infestation rate of 10.46%, judged to be above the tolerance threshold established by Colin [8], which is 5%. This result is directly related to the total number of bees in the apiary, which shows that dense colonies prevent the parasite from overgrowth.

Table 2. Estimation of the infestation rate in different hives and batches after inhalation treatment with sage essential oil.

Hive	Infestation Rate after tr: TI%
Witness batch1	26.4
Batch2	6.09
Batch3	2.32
Batch4	0.9
Hives treated/Essential oil	10.46
Chemically treated batch	9.97

The highest infection rate is that of the control with 26.4%, followed by that of the batch treated with the essential oil and then that treated with the chemical (Bayvarol). This result shows the toxic effect of the oil on the parasite (10.46%), which is more important compared to that of the chemical evaluated at 9.97%.

2.2.1. Comparison between the Effect of the Different Doses of EO in the Batches 01, 02, 03, Treated with the Chemical Treatment

The comparison between the effect of the different doses of EO in the batches 01, 02, 03, treated with the chemical treatment, is noted in Figure 1.

The essential oil of sage has a toxic effect on varroa, and it is higher (10.46%) than that of chemical treatment (9.97%).

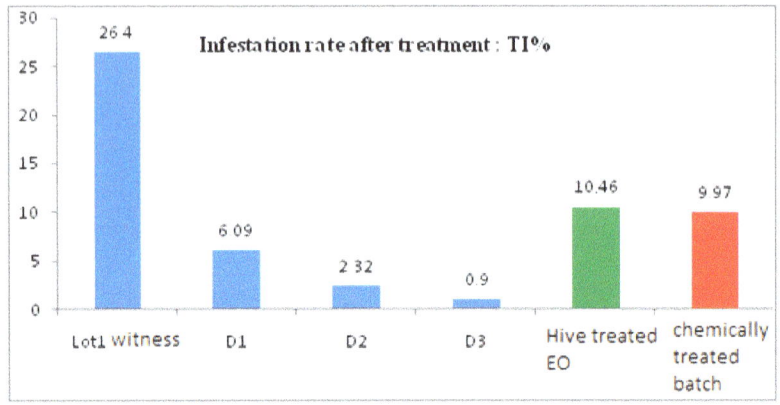

Figure 1. Infestation rate after treatment. EO: Essential Oil.

2.2.2. Statistical Analysis of the Results: Variance Analysis

- Number of varroa in the apiary (10 hives) before treatment

According to the analysis of the variance performed with the GLM test, the number of varroa that died varies marginally ($p = 0.093$, $p < 5\%$), depending on the number of bees in the hives, and does not depend on the treatment period. The most populated hives are 10 and 9 during the period of 13 November (Figure 2).

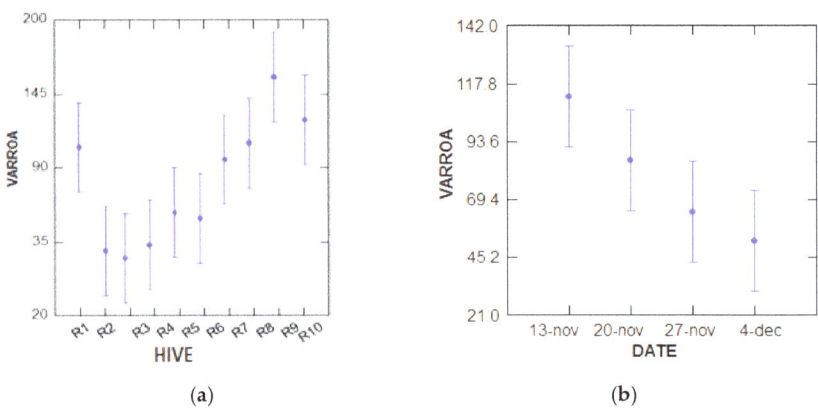

Figure 2. The variance analysis of the number of varroa that died naturally. (**a**) Hive; (**b**) Date.

- The number of varroa in the apiary after treatment

According to the variance analysis performed with the GLM test, the number of varroa that died varies in a highly significant manner ($p = 0.000$, $p < 5\%$), depending on the number of bees in the hives, and does not depend on the number of treatment period. The most populated hives are 01 and 07 during the period of 25 December (Figure 3).

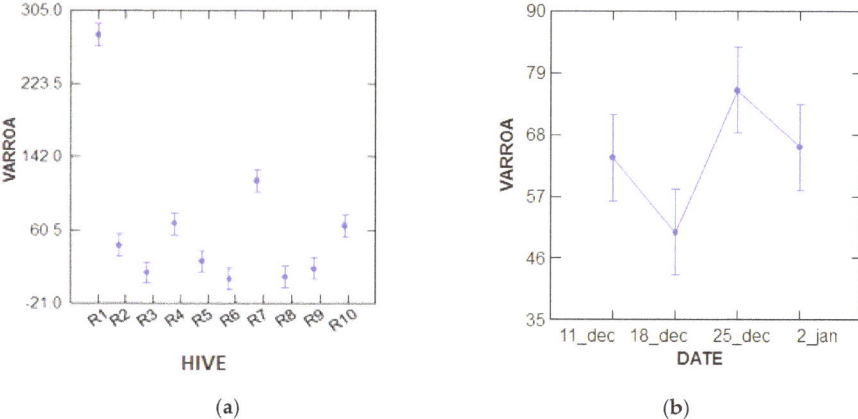

Figure 3. Analysis of the variance of the number of dead varroa by treatment. (**a**) Hive; (**b**) Date.

- Global analysis (all factors: date, hive, treatment)

According to the hive factor (colonies) and treatment:

According to the variance analysis performed with the GLM test, the number of fallen varroa varies in a highly significant manner ($p = 0.000$, $p < 5\%$) depending on the number of bees in the hives, without considering whether they are treated or not treated (Figure 4).

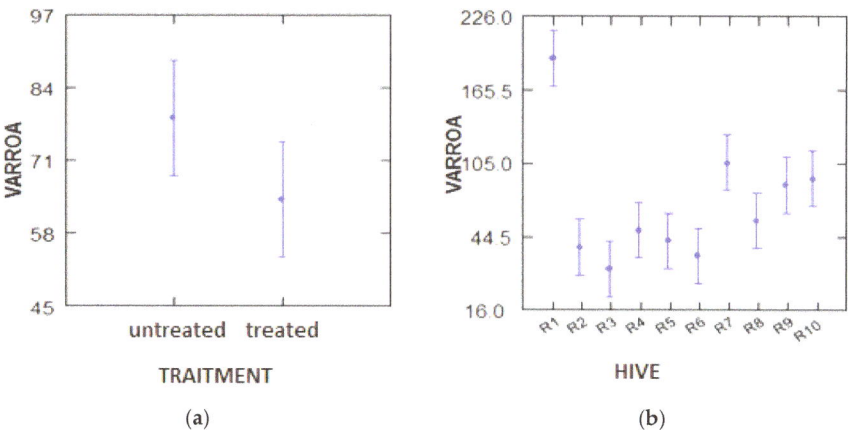

Figure 4. Analysis of the variance of the global number of dead varroa according to the hive factor and treatment. (**a**) Traitment; (**b**) Hive.

Depending on the period and treatment factor:

According to the variance analysis performed with the GLM test, the overall number of dead varroa did not depend on the treatment or the treatment period (Figure 5).

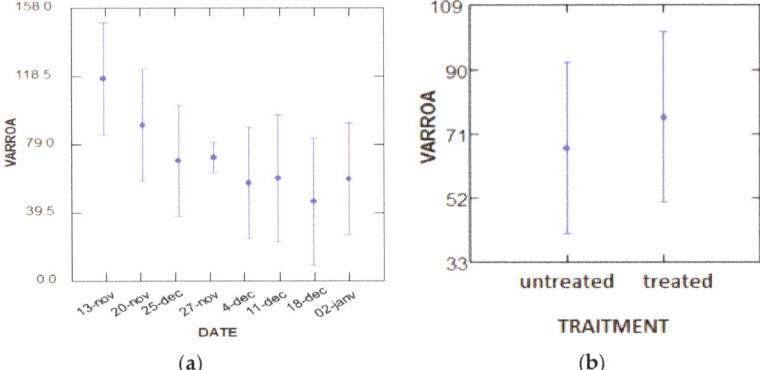

Figure 5. Analysis of the variance of the global number of dead varroa by factor period and treatment. (**a**) Date; (**b**) Traitment.

3. Discussion

Varroasis is a major problem and worrying when considering apiaries, and it has a proliferative capacity, thus causing the annihilation of the colonies of bees. To fight this enemy bee, beekeepers are oriented towards a non-pharmacopoeia reasoned with varroacids, giving birth to cases of resistance accentuated by quantities of residues which upset the products of the hive and human health. The orientation towards natural means such as essential oils of plants aromatics offers a valid solution because their presence is acceptable in the environment of the hive.

Currently, in Europe, several products are applied, and the most used are based on Fluvalinate and Amitraz. It should be noted that no treatment is shown to be 100% effective. Several studies have been conducted, reporting adverse effects of several acaricides on the health of bee colonies [16]. According to Pettis et al. [17], acaricide exposure enhances the susceptibility of bees to diseases and increases bee mortality from these diseases. Varroa mite management requires that acaricide treatments be varroa mite-selective, fatal to varroa mites at doses that are harmless to bees, and leave no or minimal residues in honey and wax [18].

Plants are capable of producing a wide variety of natural substances. In fact, in addition to the classic primary metabolites (carbohydrates, proteins, lipids, nucleic acids), they synthesize and accumulate perpetually secondary metabolites whose physiological function is not always obvious but which represent an immense source of molecules that can be exploited in various fields, among others, phytoprotection [19].

Currently, essential oils are beginning to generate much interest as a potential source of bioactive natural molecules. These products are being studied for their possible use as an alternative to insecticides, acaricides, bactericides, nematicides, and fungicides [20].

In this work, we study the acaricidal activity of *Salvia officinalis* essential oil, which has not been studied so far and comes from the Algiers region, extracted from the leaves of the plant by the hydro-distillation method. The latter allowed us to recover an essential oil yield of 0.56%, which is very low and can be explained by the influence of the region (origin), the sampling period, the climatic conditions, and the method of the extraction.

Our results from the anti-mite treatment reveal an effect of the acaricidal activity of the essential oil of *Salvia officinalis* L. on the parasitic *Varroa destructor* of the honey bee *Apis mellifera intermissa*. This acaricidal activity varies with the dose and the period of exposure to treatment. After treatment, we found that the mortality rate presents a better result when using the dose D1: 5% and that corresponds to 21.07%. This result is weak compared to that of Ghomari et al. [21], who obtained 48.7.20% of the essential oil of *Origanum vulgare*, higher than that obtained by Apivar (3.13%), and much better than that of *Thymus vulgaris*, described as being disappointing according to Giovenazzol et al. [22]. Moussaoui et al. [23] have shown that the toxicity of the *Eucalyptus* bioproduct occurs early.

The bee is an excellent biological indicator. It signals the state of health of the environment in which it lives. It detects the presence of phytosanitary substances, pollutants such as heavy metals, and radionuclides. It also ensures biodiversity through its role as a pollinator. The bee deserves to be protected.

4. Materials and Methods

4.1. Plant Material

The plant used in this study is the spontaneous sage *Salvia officinalis* (Figure 6). In our study, 30-cm apical branches were collected from Chrea, a mountainous area named Tell Atlas, near Blida, Northern Algeria (Latitude: 36°25′32″ N, Longitude: 2°52′36″ E, Altitude 946m), just before the appearance of the first floral bud in March 2016.This geographical location offers the plant a typically Mediterranean climate.

Figure 6. *Salvia officinalis* collected.

They were dried at room temperature and stored in paper bags according of the method of Branislava et al. [24]. The identification of this plant was confirmed according to the general herbarium package available at the High National School of Agronomy of El-Harrach (Algiers).

4.2. Animal Material

The *V. jacobsoni* O., 1904 (Arthropod: Dermanissidae or Varroidae) parasite of the domestic honeybee *A. mellifera intermissa* was discovered for the first time on the island Indonesian archipelago on *Apis cerana* (*A. indica*) by the entomologist Jacobson, but his study and description was made by Oudemans in 1904.

This parasite was identified according to the observations of the entomologist teachers at the High National School of Agronomy of El-Harrach—Algiers and Blida University.

Our study was conducted at the Experimental Station of the Department of Biotechnology, Faculty of Natural Sciences and Life, Blida University I. The apiary has ten hives installed in an orchard consisting of orange trees surrounded by Eucalyptus trees and those of Casuarina (Figure 7).

Figure 7. Location of the experimental apiary.

4.3. Method of Extraction of Essential Oil HS from Sage

The freshly harvested plant material was dried with a dryer at a temperature of 37° for 12 h. The aerial parts were cut into small pieces and weighed using a precision scale.

The EO was extracted by the Hydro-distillation method according to the standard procedure reported in the Sixth edition of the European Pharmacopoeia [25], using a Clevenger Type apparatus. This method involves directly immersing the plant material to be treated in a still of distilled water, which is then brought to a boil. The heterogeneous vapors are condensed on a cold surface and the essential oil separates by the difference in density [26]. After 3 h of boiling, the emerged oil is recovered in Eppendorf's. Distillation is repeated several times with 40g samples.

The yields of essential oils were expressed relative to the dry matter, according to the following formula:

$$R\% = (V/M) \times 100$$

R: percentage of the essential oil.
V: volume obtained in essential oil.
M: weight of the dry material (g).
Preparation of the dilutions
Preparation of 03 dilutions of essential oils to be tested by diluting:
D1: 0.5 mL of essential oils in 100 mL of twin.
D2: 1.5 mL of essential oils in 100 mL of twin.
D3: 02 mL of essential oils in 100 mL of twin.

Then, we prepared strips of blotting paper 18cm long and 5cm wide, each impregnated with 5mL of the different dilutions (D1, D2, D3).

For chemical treatment with Bayvarol, we used two strips per hive that were placed vertically between the frames (Figure 8).

Figure 8. Treatment of colonies with Bayvarol.

4.4. Method for Estimating the Initial Infestation Rate of Different Hives

We applied one of the biological methods "laying nappies", using diapers on greased leaves placed on the floor of 10 hives (Figure 9).

The number of varroa was counted weekly for one month before treatment and one month after treatment.

We thus estimated, by a simple division, the mortality.

The daily mortality estimate was made by dividing the total number of Varroa by 29 days; this value was multiplied by 90 days (the maximum life span of Varroa females). This allowed us to obtain the approximate number of existing Varroa in in the colony [26].

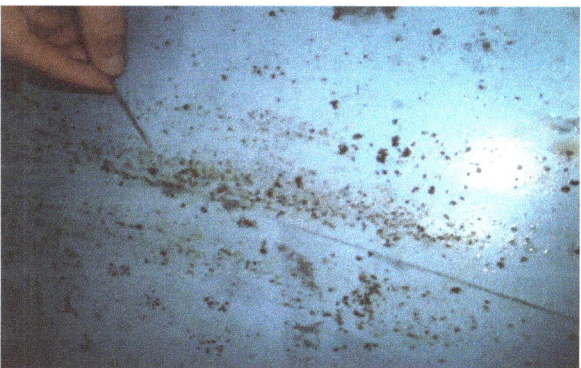

Figure 9. Counting varroa.

4.5. Method for Estimating the Number of Bees in a Colony

It was easy for us to estimate the number of bees in our hives, because a Langstroth frame contains 250 grams of bees and the average weight of a bee is estimated at 0.1 g, so a frame would have 2500 bees (250/0.1 = 2500) [27].

4.6. Method for Calculating Colony Infestation Rate

After estimating the number of Varroa and bees in a colony, the infestation rate of this colony was estimated as follows:

$$T \cdot I = C/P$$

C: corresponds to the number of varroa estimated in a colony.
P: corresponds to the number of bees estimated in a colony [15].

4.7. Method for Estimating the Infestation Rate of Different Hives after Treatment

We applied the same method as previously used.

4.8. Method for Calculating Colony Infestation Rate

After estimating the number of Varroa that fell after the application of the treatment in the colony, the infestation rate of the colony was evaluated as follows:

$$TF\% = C - M/P$$

C: corresponds to the number of varroa estimated in the colony before treatment.
M: corresponds to the number of varroa that fell after the treatment.
P: corresponds to the number of bees estimated in a colony [15].

4.9. Method for Studying the Phytotoxic Activity of Sage

The essential oil of sage was tested for its phytotoxic activity, at different doses, to obtain the resistance of *Apis mellifera* bees.

In glass jars, five individuals of *A. mellifera* bees were placed on absorbent paper; each beaker contained one dose of sage essential oil (5%, 15%, and 20%) and was covered with a pierced tissue to ensure the bees' breathing.

We found that the essential oil of sage is active in a toxic way against bees, as follows:

The essential oil of a 5% dose: more than 30 min to notice the beginning of the death of some bees.
The essential oil of a 15% dose: 14.34 min to notice the death of almost all the bees.
The essential oil of a 20% dose: 6.25 min to notice the death of all the bees.

5. Conclusions

A preliminary diagnosis can be made after opening brood cells and observations of immature and adult mites present in them or by the biological method "laying nappies", which reveals, in our study, an initial infestation rate of 11.53%. This rate is close to the range 10 and 20% according to Robaux [15], which means that the colony is strongly affected and requires treatment. This treatment can be done at the level of the hive with various chemicals not without danger since it destroys the mites with negative effects on: the bee, the frames and supports, and the honey. But our study is part of the simple and economical biological method to treat Varroa on the one hand, and collect and observe hive debris and mites for scientific purposes, on the other hand. Itis the method of fumigation treatment using *Salvia officinalis*. It is clear that treatment with Chrea sage reduced the final infestation rate to 10.46, and for the untreated lot, it was 26.4%. Also, statistical analyzes showed that sage essential oil treatments have a significant effect. However, the weakness of the effectiveness of the treatment has its origin in the presence of the capped broods which "protect" the varroa inside the alveoli and thus prevent the penetration of the smoke. In other words, the varroa attached to the lower part of the body of the larva escape, unfortunately, the effects of treatment. Thus, it becomes imperative in our view for beekeepers to ensure the state of the hive before the period of the slope of the eggs to avoid any contamination.

We conclude that the effectiveness of essential oils is related to the plant species, the dose used, and the duration of exposure.

Author Contributions: R.B. and L.H. conceived and designed the experiments; L.B. performed the experiments; K.A. analyzed the data; R.B. contributed reagents/materials/analysis tools; L.B. wrote the paper.

Funding: The founding sponsors had no role in the design of the study; in the collection, analyses, or interpretation of data; in the writing of the manuscript, and in the decision to publish the results.

Acknowledgments: This study is part of doctoral research. The authors thank all those who contributed to the achievement of this work, including the head of the experimental station of the Department of Biotechnology, Faculty of Natural Sciences and Life, Blida University I. We also thank the responsible for the herbarium at the High National School of Agronomy of El-Harrach—Algiers. The authors are grateful to Sadek Oudni, Djouhra Khitouche, and Rima Zedadra for their help. Last but not least, we thank Pr. Kamal Bechkoum for reading the manuscript and sending us helpful comments from Gloucestershire University in the UK, and the reviewers of the manuscript for their comments and suggestions.

Conflicts of Interest: The authors declare no conflict of interest.

References

1. Clement, H. *L'abeille Sentinelle de L'environnement*; Editions Alternatives: Paris, France, 2009.
2. Garcia-Fernandez, P.; Rodriguez, R.B.; Orantes-Bermejo, F.J. Influence du climat sur le développement de la population de *Varroa jacobsoni* Oud. dans des colonies d'*Apis mellifera iberica* (Goetze) dans le sud de l'Espagne. *Apidologie* **1995**, *26*, 371–380. [CrossRef]
3. Ellis, J.D.; Zettel Nalen, C.M. *Varroa Mite, Varroa destructor Anderson and Trueman (Arachnida: Acari: Varroidae)*; EENY-473; IFAS Extention, University of Florida: Gainesville, FL, USA, 2010.
4. Colin, M.E. La varroase. *Rev. Sci. Tech.* **1982**, *1*, 1177–1189. [CrossRef]
5. Belaid, M.; Doumandji, S. Effet du varroa distructor sur la morphométrie allaire et sur les composants du systhème immunitaire de l'abeille ouvrière *Apis mellifera intermisa*. *Leban. Sci. J.* **2010**, *11*, 45–53.
6. Teffahi, M.; Belguendouz, R. Etude de L'activité Acaricide de L'origan (Origanum glandulosum) et du Romarin (Rosmarinus officinalis) Sur le Parasite (Varroa jacopsoni) de L'abeille. Master's Thesis, Blida University, Blida, Algeria, 2014.
7. Ghorbani, A.; Esmaeilizadeh, M. Pharmacological properties of *Salvia officinalis* and its components. *J. Tradit. Complement. Med.* **2017**, *7*, 433–440. [CrossRef] [PubMed]
8. Colin, M.E. A method for characterizing the biological activity of Essential oils against *Varroa jacobsoni*. In *New Perspectives on Varroa A Matheson*; IBRA: Cardiff, UK, 1994.
9. Badiee, P.; Nasirzadeh, A.R.; Motaffaf, M. Comparison of *Salvia officinalis* L. essential oil and antifungal agents against *candida* species. *J. Pharm. Technol. Drug. Res.* **2012**. [CrossRef]
10. Hayouni, E.A.; Chraief, I.; Abedrabba, M. Tunisian *Salvia officinalis* L. and *Schinusmolle* L. essential oils: their chemical compositions and their preservative effects against *Salmonella* inoculated in minced beef meat. *Int. J. Food Microbiol.* **2008**, *125*, 242–251. [CrossRef] [PubMed]
11. Langer, R.; Mechtler, C.; Jurenitsch, J. Composition of the essential oils of commercial samples of *Salvia officinalis* L. and *S. fruticosa* Miller: A comparison of oils obtained by extraction and steam distillation. *Phytochem. Anal.* **1996**, *7*, 289–293. [CrossRef]
12. Mills, S.; Bone, K. *The Essential Guide to Herbal Safety*; Elsevier: Louis, MO, USA, 2005.
13. Veličković, D.T.; Ranđelović, N.V.; Ristić, M.S.; Veličković, A.S.; Šmelcerović, A.A. Chemical constituents and antimicrobial activity of the ethanol extracts obtained from the flower, leaf and stem of *Salvia officinalis* L. *J. Serb. Chem. Soc.* **2003**, *68*, 17–24. [CrossRef]
14. Russo, A.; Formisano, C.; Rigano, D. Chemical composition and anticancer activity of essential oils of Mediterranean sage (*Salvia officinalis* L.) grown in different environmental conditions. *Food Chem. Toxicol.* **2013**, *55*, 42–47. [CrossRef] [PubMed]
15. Robaux, P. *Varroa et Varroatose*; Opida: Echauffour, France, 1986.
16. Pettis, J.S. A scientific note on Varroa destructor resistance to coumaphos in the United States. *Apidologie* **2004**, *35*, 91–92. [CrossRef]

17. Pettis, J.S.; Lichtenberg, E.M.; Andree, M.; Stitzinger, J.; Rose, R.; Vanengelsdorp, D. Crop pollination exposes honey bees to pesticides which alters their susceptibility to the gut pathogen Nosema ceranae. *PLoS ONE* **2013**, *8*. [CrossRef] [PubMed]
18. Lindberg, C.M.; Melathopoulos, A.P.; Winston, M.L. Laboratory evaluation of miticides to control Varroa jacobsoni (Acari: Varroidae), a honey bee (Hymenoptera: Apidae) parasite. *J. Econ. Entomol.* **2000**, *93*, 189–198. [CrossRef] [PubMed]
19. Auger, J.; Thibout, E. Susbtances soufrées des Allium et des Crucifères et leurs potentialités phytosanitaires. In *Biopesticides D'origine Végétale*; Regnault-Roger, C., Philogène, B.J.R., Vincent, C., Eds.; Lavoisier, Tec & Doc: Paris, France, 2002.
20. Yakhlef, G. Etude de L'activité Biologique des Extraits de Feuilles de *Thymus vulgaris* L. et *Laurius nobilis* L. Master's Thesis, EL Hadj Lakhdar—Batna University, Batna, Algeria, 2010.
21. Ghomari, F.N.; Kouache, B.; Arous, A.; Cherchali, S. Effet de traitement par fumigation du thym (*Thymus vulgaris*) sur le *Varroa destructor* agent de la varroase des abeilles. *Nat. Technol.* **2014**, *10*, 34–38.
22. Giovenazzol, P.; Marceau, J.; Dube, S. Essais préliminaires sur le traitement de colonies d'abeilles *Apis mellifera* infestées par le parasite *Varroa jacobsoni* en chambre d'hivernage. *L'abeille* **1999**, *19*, 5. Available online: www.agrireseau.net/apiculture/Documents/essais_traitement_acariose.pdf (accessed on 5 June 2018).
23. Moussaoui, K.; Ahmed Hedjala, O.; Zitouni, G.; Djazouli, Z. Estimation de la toxicité des d'huiles essentielles formulées de thym et d'eucalyptus et d'un produit de synthèse sur le parasite de l'abeille tellienne varroa destructor (arachnida, varroidae). *Agrobiologie* **2014**, *4*, 17–26.
24. Branislava, S.; Lakušić, C.; Mihailo, S.; Ristić, A.; Violeta, N.; Slavkovska Danilo, L.; Stojanović, C.; Dmitar, V.; Lakušić, C. Variations in essential oil yields and compositions of *Salvia officinalis* (Lamiaceae) at different developmental stages. *Bot. Serb.* **2013**, *37*, 127–139.
25. European Pharmacopoeia Commission. *Ph. Eur. 6.0. Council of Europe*; European Pharmacopoeia Commission: Strasbourg, France, 2007.
26. Bruneton, J. *Pharmacognosie, Phytochimie, Plantes Médicinales*, 3rd ed.; Technique et DocumentationLavoisier: Paris, France, 1999.
27. Berkani, M.L. Comparaison de Deux Types de Ruches: Dadant et Langstroth Dans les Littoral Est et Algérois. Master's Thesis, High school of Agronomy, El Harrach, Algeria, 1985.

© 2018 by the authors. Licensee MDPI, Basel, Switzerland. This article is an open access article distributed under the terms and conditions of the Creative Commons Attribution (CC BY) license (http://creativecommons.org/licenses/by/4.0/).

Article

Polar Constituents of *Salvia willeana* (Holmboe) Hedge, Growing Wild in Cyprus

Theofilos Mailis and Helen Skaltsa *

Department of Pharmacognosy & Chemistry of Natural Products, School of Pharmacy, National and Kapodistrian University of Athens, Panepistimiopolis, Zografou, 157 71 Athens, Greece; skaltsa@pharm.uoa.gr
* Correspondence: skaltsa@pharm.uoa.gr; Tel.: +30-21-0727-4593

Received: 29 January 2018; Accepted: 1 March 2018; Published: 6 March 2018

Abstract: Twenty compounds were isolated from the aerial parts of *Salvia willeana* (Holmboe) Hedge, growing wild in Cyprus. These compounds comprise one new and one known acetophenone, one megastigmane glucoside, five phenolic derivatives, two caffeic acid oligomers, three flavonoids, two lignans, two triterpene acids, one monoterpene glucoside, and two fatty acids. The structures of the isolated compounds were established by means of NMR [(Rotating-frame OverhauserEffect SpectroscopY) (^{1}H-^{1}H-COSY (COrrelation SpectroscopY), ^{1}H-^{13}C-HSQC (Heteronuclear Single Quantum Correlation), HMBC (Heteronuclear Multiple Bond Correlation), NOESY (Nuclear Overhauser Effect SpectroscopY), ROESY (Rotating-frame Overhauser Effect SpectroscopY)] and MS spectral analyses. This is the first report of the natural occurrence of 4-hydroxy-acetophenone 4-*O*-(3,5-dimethoxy-4-hydroxybenzoyl)-*β*-D-glucopyranoside. A chemical review on the non-volatile secondary metabolites has been carried out. Based on the literature data, the analysis revealed that the chemical profile of *S. willeana* is close to that of *S. officinalis* L.

Keywords: *Salvia willeana*; 4-hydroxy-acetophenone 4-*O*-(3,5-dimethoxy-4-hydroxybenzoyl)-*β*-D-glucopyranoside; megastigmane glucoside; phenolics; terpenes; 2D NMR; *Salvia* L: chemical review

1. Introduction

The Lamiaceae family consists of more than 250 genera; *Salvia* L. is the largest genus within this family due to the presence of approximately 900 species. *Salvia* L. spreads in the warm and temperate regions of both the northern and southern hemispheres, and some species of this genus have been cultivated worldwide for use in folk medicine, in perfumery and cosmetics industries, as well as for culinary purposes, like flavoring and aromatic agents [1,2]. Some of its many interesting biological and pharmacological properties are its antioxidant [3], antimicrobial [3,4], cytotoxic [3,4], anti-HIV [3], and antiplasmodial effects [4], as well as others [3]. It is noteworthy that the name of the genus, *Salvia*, is derived from the Latin word "salvare", which means "to save", in reference to the curative properties of the plant [5]. The genus has attracted such great interest, that it has become the subject of numerous chemical studies, giving evidence that these plants are a rich source of a wide variety of secondary metabolites, such as polyphenols and terpenoids [1].

Salvia willeana (syn. *S. grandiflora* subsp. *willeana* Holmboe and *S. grandiflora* subsp. *albiflora* Lindb.) is a low-growing, strongly aromatic suffruticose herb, sometimes carpeting the ground [6]. This species is endemic to Cyprus, where it grows on moist, rocky mountainsides of the Troodos range at 1150–1950 m altitude and it flowers from May to October [7]. Its extracts possess different pharmacological properties and the plant has been used to halt milk production in nursing mothers, as well as for its antiseptic activity [8]. As *S. willeana* is locally used in aqueous preparations, the purpose of our study was the investigation of secondary metabolites obtained from the polar extract of its aerial parts. Our previous study of its lipophilic extract, revealed the presence of camphor, lupeol, and oleanolic acid, and demonstrated their anti-inflammatory effect [8]. It is of interest to note that there is only one more report regarding the chemical constituents of a polar extract of *S. willeana*, which revealed the presence of the triterpenoids ursolic and oleanolic acids, the diterpenoids carnosic acid and isorosmanol, and the flavonoid salvigenin [9].

2. Results

The polar extract (MeOH:H$_2$O 5:1) of *S. willeana* was fractionated by MPLC (medium pressure liquid chromatography), CC (column chromatography), followed by semi-preparative HPLC, preparative TLC, and yielded two triterpenoids, namely ursolic acid (**1**) and maslinic acid (**2**), one monoterpene glucoside, (1*S*,2*R*,4*R*)-1,8-epoxy-*p*-menthan-2-yl-*O*-β-D-glucopyranoside (**3**), one megastigmane glucoside, (6*R*,9*S*)-3-oxo-α-ionol β-D-glucopyranoside (**4**), five phenolic compounds, i.e., hydroxy-tyrosol (**5**), *p*-anisic acid (**6**), eleutheroside B (syringin) (**7**), 1-*O*-*p*-hydroxybenzoyl-β-D-apiofuranosyl-(1→6)-β-D-glucopyranoside (**8**) and eugenylglucoside (**9**), two acetophenones, 4-*O*-β-D-glucopyranosyl acetophenone (picein) (**10**) and 4-hydroxy-acetophenone 4-*O*-(3,5-dimethoxy-4-hydroxybenzoyl)-β-D-glucopyranoside (**11**), two caffeic acid oligomers, rosmarinic acid (**12**) and salvianolic acid K (**13**), three flavonoids, luteolin-7-*O*-β-D-glucoside (cynaroside) (**14**), 6-hydroxyluteolin 7-*O*-β-D-glucoside (**15**) and hesperidin (**16**), two lignans, syringaresinol-4-*O*-β-D-glucopyranoside (**17**), pinoresinol-4-*O*-β-D-glucopyranoside (**18**), and two fatty acids: linoleic acid (**19**) and methyl α-linolenate (**20**) (Figure 1).

Compound **11** was obtained as a white amorphous powder. $[\alpha]_D^{20}$ − 4.71° (c 0.25 MeOH); UV (CH$_3$OH) λ_{max}: 272 nm. The HREIMS of **11** established its molecular formula as C$_{23}$H$_{26}$O$_{11}$ (found 477.1405 [M − H]$^-$, calcd. 478.1449). The ^{1}H-NMR spectrum (Table 1) showed signals at 7.69 (2H, d, *J* = 9.0 Hz) and 7.05 (2H, d, *J* = 9.0 Hz), which were indicative of a 1,4-bisubsituted phenyl group. Thus, these proton signals were recognized belonging to the aromatic ring of the acetophenone moiety [10]. In the upfield region of the ^{1}H-NMR spectrum, a singlet at δ_H 2.47 (3H, s) was ascribed to the methyl group (CH$_3$-8) attached on the carbonyl group of the acetophenone [10]. Moreover, the presence of a singlet at δ_H 7.33 with an integration of two aromatic protons, indicative of a pair of equivalent methine protons, revealed the occurrence of the syringic acid ester structure in the molecule [11]. In the ^{1}H-NMR spectrum, the presence of a β-D-glucopyranose unit was evident based on a characteristic doublet signal with a coupling constant of 7.9 Hz at δ_H 5.04, assignable to the anomeric proton of the sugar moiety. Moreover, in the ^{1}H-^{1}H-COSY spectrum the correlation peaks between the vicinal protons of the sugar ring were observed. Furthermore, from the ^{13}C-NMR data (Table 1) the carbon signals of the glucose moiety were assigned at δ_C 101.7 (C-1′), 77.5 (C-3′), 75.5 (C-5′), 74.5 (C-2′), 71.3 (C-4′), and 64.8 (C-6′), matching the reported data of 1,6-disubstituted-β-D-glucose [12]. The structural assignment was further confirmed by HSQC experiments, due to the carbon signals at δ_C 130.1 (C-2/C-6), 116.2 (C-3/C-5), and 26.5 (C-8), indicative of a 4-hydroxy-acetophenone moiety [9,13], while the syringic acid ester was confirmed by the presence of the carbon signals at δ_C 108.3 (C-2″/C-6″), 56.0 (3″, 5″-OCH$_3$) [11]. The existence of the acetophenone moiety was corroborated by the HMBC experiment. This spectrum revealed a long-range cross peak between the carbonyl group at δ_C 199.1 (C-7) with the equivalent protons H-2/H-6 (δ_H 7.69), as well as an interaction between the carbonyl group (δ_C 199.1) and the methyl group CH$_3$-8 (δ_H 2.47). In addition, the linkage of the syringyl moiety with the glucosyl moiety was substantiated by the observation of an HMBC correlation between the carbonyl group at δ_C 167.3 and the 6′-methylene protons at δ_H 4.46 (dd, *J* = 11.7, 8.0). Furthermore, the position of the attachment of the carboxyl group to the quaternary carbon C-1″ of the syringic ester was determined by a diagnostic HMBC cross peak between the equivalent methine protons H-2″/H-6″ at δ_H 7.33 and the carbonyl carbon C-7″ at δ_C 167.3, while a long-range coupling between the methyl protons of the methoxy groups at δ_H 3.83 and the benzylic carbons C-3″/C-5″ at δ_C 149.0 was also observed. Moreover, the position of the attachment of glucose to the 4-hydroxy-acetophenone moiety was revealed by a ROESY experiment, which displayed correlations between the anomeric proton H-1′ (δ_H 5.04) and the equivalent protons H-3/H-5 (δ_H 7.05) of the acetophenone (Figure 2). On the basis of the information above and by comparison with the data for compounds of similar structures [10,11,13] compound **11** was identified as 4-hydroxy-acetophenone 4-*O*-(3,5-dimethoxy-4-hydroxybenzoyl)-β-D-glucopyranoside, which is a new natural product, to the best of our knowledge.

The identification of the known flavonoids luteolin-7-*O*-β-D-glucoside (**14**) [14–16], 6-hydroxyluteolin 7-*O*-β-D-glucoside (**15**) [17], and hesperidin (**16**) [18–24] was based on UV–VIS and NMR spectroscopic analyses, as well as by comparing their spectroscopic data with those reported in the literature. The structure of the two fatty acids, linoleic acid (**19**) [25,26] and methyl α-linolenate (**20**) [26,27], has been deduced by the interpretation of NMR and GC-MS data. The ^{1}H-NMR chemical shifts for compounds **1** [28–30], **2** [31–33], **3** [34,35], **4** [35–37], **7** [38–42], **9** [35,43], **10** [10,35,44], **12** [35,45–48], **17** [49,50], and **18** [51–54] presented in our study are in agreement with the data previously reported in the literature.

However, the NMR data of compounds **5**, **6**, **8**, and **13** are not fully recorded in the literature, therefore, they are presented here below.

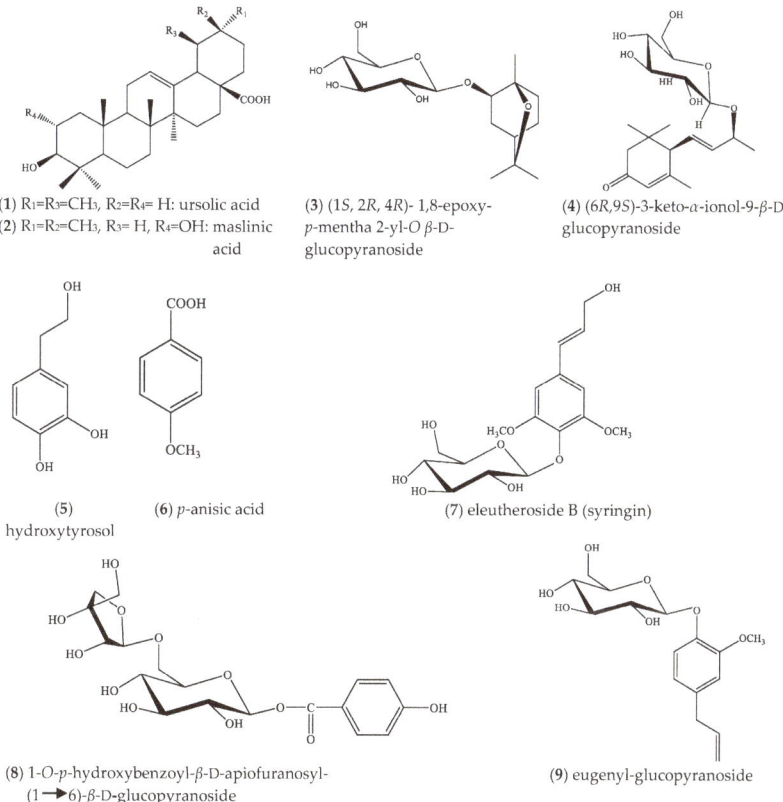

(**1**) R$_1$=R$_3$=CH$_3$, R$_2$=R$_4$= H: ursolic acid
(**2**) R$_1$=R$_2$=CH$_3$, R$_3$= H, R$_4$=OH: maslinic acid

(**3**) (1*S*, 2*R*, 4*R*)- 1,8-epoxy-*p*-mentha 2-yl-*O* β-D-glucopyranoside

(**4**) (6*R*,9*S*)-3-keto-α-ionol-9-β-D-glucopyranoside

(**5**) hydroxytyrosol

(**6**) *p*-anisic acid

(**7**) eleutheroside B (syringin)

(**8**) 1-*O*-*p*-hydroxybenzoyl-β-D-apiofuranosyl-(1→6)-β-D-glucopyranoside

(**9**) eugenyl-glucopyranoside

Figure 1. *Cont.*

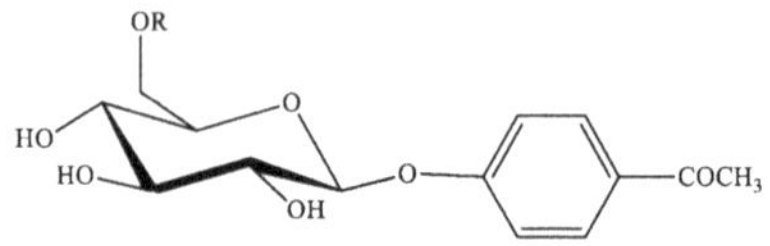

(10)	R= H		4-O-β-D-glucopyranosyl acetophenone (picein)
(11)	R=	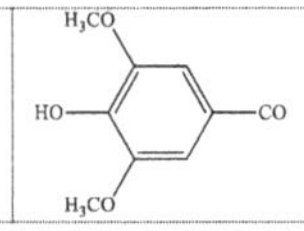 (3,5-dimethoxy-4-hydroxybenzoyl group)	4-hydroxy-acetophen-4-O-(3,5-dimethoxy-4-hydroxybenzoyl)-β-D-glucopyranoside

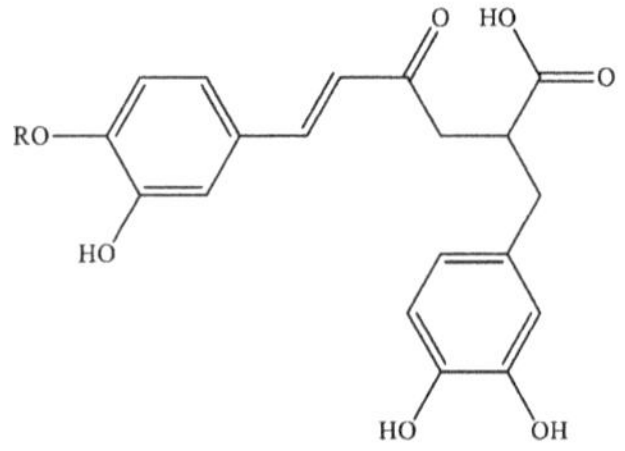

(12)	R=	H	rosmarinic acid
(13)	R=	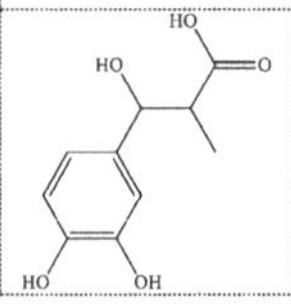 (structure)	salvianolic acid K

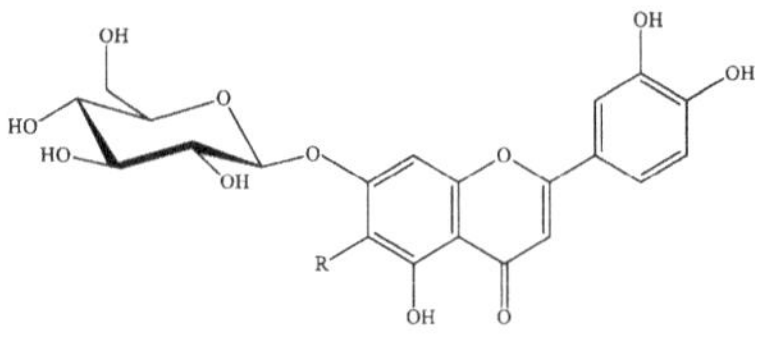

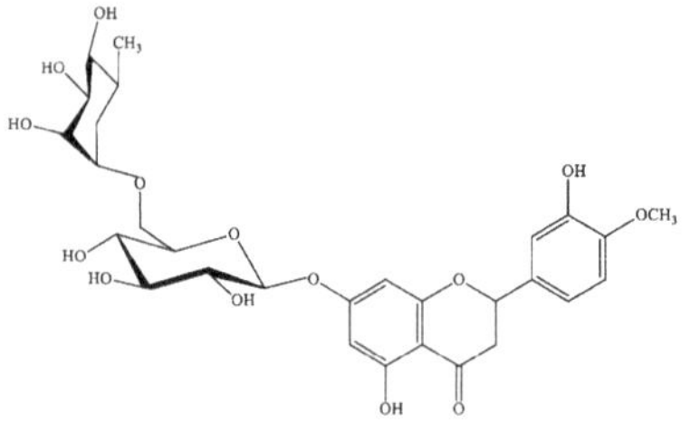

(14): R=H luteolin-7-O-β-D-glucoside (cynaroside)
(15): R=OH 6-hydroxy-luteolin-7-O-β-D-glucopyranoside

(16) hesperidin

Figure 1. Cont.

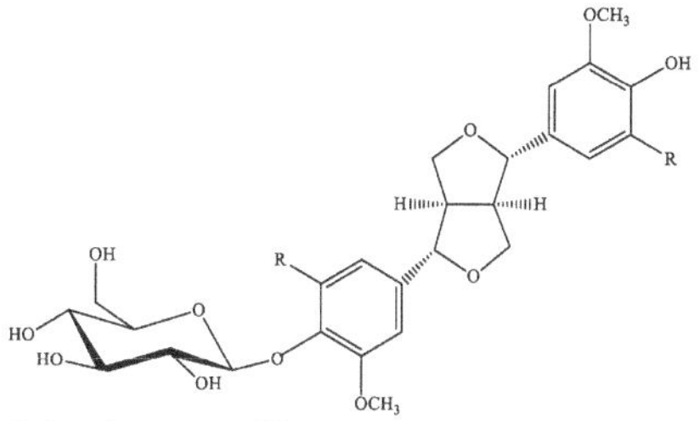

(17) R= CH₃: syrigaresinol-4-*O*-β-D-glucopyranoside
(18) R=H: pinoresinol-4-*O*-β-D-glucopyranoside

(19) linoleic acid (C18:2)

(20) methyl α-linolenate (C18:3)

Figure 1. Structures of isolated compounds from *Salvia willeana*.

Table 1. ^{1}H-NMR and ^{13}C-NMR spectrum of **11**.

	δ_C	C	δ_H	H	J (Hz)
1	131.5	C	-	-	-
2	130.1	CH	7.69	1	d (J = 9.0)
3	116.2	CH	7.05	1	d (J = 9.0)
4	162.5	C	-	-	-
5	116.2	CH	7.05	1	d (J = 9.0)
6	130.1	CH	7.69	1	d (J = 9.0)
7	199.1	C	-	-	-
8	26.5	CH$_3$	2.47	3	s
1′	101.7	CH	5.04	1	d (J = 7.8)
2′	74.5	CH	3.53	2	m
3′	77.5	CH			
4′	71.3	CH	3.43	1	m
5′	75.5	CH	3.89	1	dd (J = 8.0, 2.3)
6a′	64.8	CH$_2$	4.71	2	dd (J = 11.7, 2.3)
6b′			4.46		dd (J = 11.7, 8.0)
1″	-	C	-	-	-
2″	108.3	CH	7.33	1	s
3″	149.0	C	-	-	-
4″	142.3	C	-	-	-
5″	149.0	C	-	-	-
6″	108.3	CH	7.33	1	s
7″	167.3	C	-	-	-
3″, 5″-OCH$_3$	56.0	CH$_3$	3.83	6	s

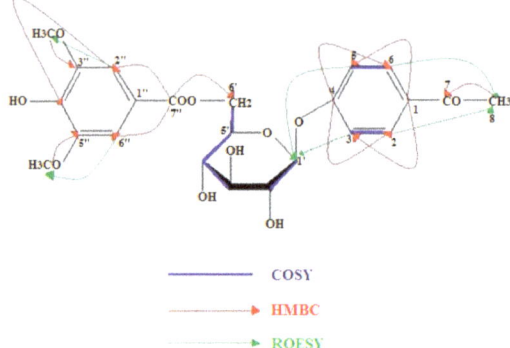

Figure 2. COSY, HMBC, and ROE signals of compund **11**.

3. Discussion

The genus *Salvia* L. is characterized by the presence of several different secondary metabolites, mainly phenolic derivatives and terpenoids [3].

In the present study, overall, 20 compounds were isolated from *S. willeana* polar extracts, i.e., two triterpenoids, namely ursolic acid (**1**) and maslinic acid (**2**), one monoterpene glucoside, (1*S*,2*R*,4*R*)-1,8-epoxy-*p*-menthan-2-yl-*O*-β-D-glucopyranoside (**3**), one megastigmane glucoside, (6*R*,9*S*)-3-oxo-α-ionol β-D-glucopyranoside (**4**), five simple phenolic compounds, i.e., hydroxy-tyrosol (**5**), *p*-anisic acid (**6**), eleutheroside B (syringin) (**7**), 1-*O*-*p*-hydroxybenzoyl-β-D-apiofuranosyl-(1→6)-β-D-glucopyranoside (**8**) and eugenylglucoside (**9**), two acetophenones, 4-*O*-β-D-glucopyranosyl acetophenone (picein) (**10**) and 4-hydroxy-acetophenone 4-*O*-(3,5-dimethoxy-4-hydroxybenzoyl)-β-D-glucopyranoside (**11**), two caffeic acid oligomers, rosmarinic acid (**12**) and salvianolic acid K (**13**), three flavonoids, luteolin-7-*O*-β-D-glucoside (cynaroside) (**14**), 6-hydroxyluteolin 7-*O*-β-D-glucoside (**15**) and hesperidin (**16**), two lignans, syringaresinol-4-*O*-β-D-glucopyranoside (**17**), pinoresinol-4-*O*-β-D-glucopyranoside (**18**) and two fatty acids: linoleic acid (**19**) and methyl α-linolenate (**20**).

It is interesting to point out that compounds **7**, **16**, **17**, and **18** had not been previously detected in *Salvia* L. Syringin (**7**) is reported, here, as a component of the Lamiaceae family for the first time. Moreover, compounds **3**, **5**, **6**, and **8–10** had been previously mentioned only once in the genus, as follows: **3**, **9**, **10** [35], **6** [55], **8** [10] and **14**, **15** [3] in *S. officinalis* L., **5** in *S. digitaloides* Diels [56], while compound **4** twice in *S. nemorosa* and *S. officinalis* L. [35,37]. As for the two triterpenoids **1** and **2**, these have previously been isolated from the acetone extract of the aerial parts of *S. willeana* [9]. So far, only 134 *Salvia* species of the over 1000 species suggested have been investigated [3]. Based on our results, concerning the polar secondary metabolites, among these species *S. willeana* showed many similarities to *S. officinalis* L., since most of the isolated simple phenols (**5**, **6**, **8–10**), as well as the flavonoids **14** and **15**, are found only in these two species (Supplementary Materials Table S1 and the references herein).

4. Materials and Methods

4.1. Plant Material

Aerial parts of *Salvia willeana* (Holmboe) Hedge were collected on Troodos Mountain in Cyprus in April 2004 [8]. A voucher specimen has been deposited in the Agricultural Research Institute Herbarium of Nicosia [no. ARI 3213].

4.2. Equipment and Reagents

^{1}H, ^{13}C, and 2D-NMR spectra were recorded in CDCl$_3$ and CD$_3$OD on Bruker DRX 400 (399.95 MHz for ^{1}H-NMR) and Bruker AC 200 (200.13 MHz for ^{1}H-NMR and 50.3 MHz for ^{13}C-NMR) instruments at 295 K (Bruker BioSpin GmbH, Silberstetten, Germany). Chemical shifts are given in ppm (δ) and were referenced to the solvent signals at 7.24/3.31 and 77.0/49.5 ppm for ^{1}H and ^{13}C-NMR, respectively. COSY, HSQC, HMBC, NOESY, and ROESY (mixing time 950 ms) were performed using standard Bruker microprograms. High-resolution mass spectra were measured on a Q-TOF 6540 UHD (Aligent Technologies, Santa Clara, California, USA). The solvents used were of spectroscopic grade (Merck KGaA, Darmstadt, Germany). UV spectra were recorded on a Shimadzu UV-160A spectrophotometer (Shimadzu; Kyoto, Japan), according to Mabry et al. [57]. Optical rotations were determined using a Perkin-Elmer Polarimeter 341 (Perkin-Elmer, GmbH, Überlingen, Germany). GC-MS (Gas Chromatography-Mass Spectrometry) analyses were performed on a Hewlett-Packard 5973–6890 system (Palo Alto, California) operating in EI mode (70 eV) equipped with a split/splitless injector (220 °C), a split ratio 1/10, using a fused silica HP-5 MS capillary column (30 m × 0.25 mm (i.d.), film thickness: 0.25 µm) with a temperature program for HP-5 MS column from 60 °C (5 min) to 280 °C, at a rate of 4 °C/min and helium as a carrier gas at a flow rate of 1.0 mL/min. Preparative HPLC (High-Performance Liquid Chromatography) was performed using a C$_{18}$ 25 cm × 10 mm Kromasil column on a HPLC system (Jasco PU-2080; JASCO, Tokyo, Japan) equipped with an RI detector Shimadzu 10A (Shimadzu, Kyoto, Japan); flow rate: 1.0 mL/min; concentration of the samples: 3.5–7.0 mg/mL. All solvents used were of HPLC grade (Merck). MPLC (Medium Pressure Liquid Chromatography) was performed using Büchi C-615 and Büchi 688 chromatographic pump; columns: Büchi Borosilikat 3.3, (41.0 cm × 4.0 cm), flow rate: 10 mL/min; (15.0 cm × 1.5 cm) flow rate: 3 mL/min; vacuum liquid chromatography (VLC): silica gel 60H (Merck, Art. 7736) [58]. Column chromatography (CC): silica gel (Merck, Art. 9385), silica gel 60 (230–400 mesh ASTM, SDS 2050044) gradient elution with the solvent mixtures indicated in each case; Sephadex LH-20 (Pharmacia Fine Chemicals); cellulose (Avicel, Merck, Art. 2330). Preparative TLC (Thin Layer Chromatography) was performed using pre-coated silica gel 60 plates (Merck, Art. 5721). Fractionation was always monitored by TLC silica gel 60 F-254, (Merck, Art. 5554) with visualization under UV (254 and 365 nm) and spraying with vanillin-sulfuric acid reagent (vanillin Merck, Art. No. S26047 841) [59] and Neu's reagent for phenolics [60]. Analytical solvents were obtained from Panreac Quimica SA (Barcelone, Spain, Italy), while deuterated solvents were purchased from Merck, KGaA (Darmstadt, Germany). Di-phosphorus pentoxide was purchased from Chemlab, Belgium.

4.3. Extraction and Chromatography

The air-dried powdered aerial parts of S. willeana (0.43 kg) were successively extracted at room temperature with cyclohexane, dichloromethane, MeOH, and MeOH:H$_2$O (5:1) (2 L of each solvent, twice for 48 h) [8]. A portion of the latter extract (9.0 g) was fractionated on a RP$_{18}$-MPLC (41.0 × 4.0 cm) using a H$_2$O: MeOH gradient system (100% H$_2$O → 100% MeOH; steps of 10% MeOH; 50 min each; 50% MeOH: 50% EtOAc 50 min; 100% EtOAc 50 min) to yield twenty three fractions (A–V) of 500 mL each. Fraction D (598.8 mg; H$_2$O:MeOH 85:15) was similarly purified by RP$_{18}$-MPLC (15 cm × 1.5 cm) to obtain three sub-fractions (1–3). The first two sub-fractions were combined together (sub-fraction DA, 364.5 mg), subjected to a Sephadex LH-20 column and eluted with 100% methanol to afford 61 fractions combined in 13 groups (DAA–DAM). Group DAJ (4.2 mg) was identified as salvianolic acid K (**13**). Group DAC (12.9 mg) was subjected to RP$_{18}$-HPLC (RID; isocratic elution using MeOH:CH$_3$COOH 5% 30:70; flow-rate: 1.0 mL/min) and afforded 1-O-p-hydroxybenzoyl-β-D-apiofuranosyl-(1→6)-β-D-glucopyranoside (**8**) (Rt = 14.0 min, 0.5 mg) and 4-O-β-D-glucopyranosyl-acetophenone (picein) (**10**) (Rt = 17.2 min, 0.7 mg). Group DAD (9.3 mg) was purified by prep. TLC on silica gel using CHCl$_3$:MeOH:AcOH (7:1.5:1.5) as the eluent and yielded hydroxytyrosol (6.3 mg) (**5**). Fraction H (718.3 mg) was fractionated by CC on a Sephadex LH-20 (25.0 cm × 3.2 cm) using H$_2$O:MeOH (20:80 to 0:100) for gradient elution to afford luteolin 7-O-β-D-glucoside (0.7 mg) (**14**). Groups HC to HG were combined (HC'; 193.7 mg) and purified by

CC (12.2 cm × 2.2 cm) over silica gel with cyclohexane: DM:EtOAc:MeOH mixtures of increasing polarity to yield nine groups (HC'A–HC'I). Group HC'C (14.2 mg; eluted with EtOAc:MeOH 97:3 to 94:6) was purified by RP$_{18}$-HPLC (RID; MeOH; H$_2$O 40:60; flow rate: 1 mL/min) to obtain syringin (Rt = 56.4 min, 0.1 mg) (7), (1S,2R,4R)-1,8-epoxy-p-menthan-2-yl-O-β-D-glucopyranoside (Rt = 74.2 min, 0.9 mg) (3), (6R,9S)-3-oxo-α-ionol-β-D-glucopyranoside (Rt = 116.5 min, 0.5 mg) (4), eugenyl-glucoside (Rt = 130.1 min, 0.3 mg) (9). Combined groups HJ to HL (HJ'; 48.9 mg; eluted with H$_2$O:MeOH 50:50) were subjected to CC over silica gel using cyclohexane: DM:EtOAc:MeOH mixtures of increasing polarity (60 fractions). Purification of fraction HJ'E (2.1 mg; eluted with EtOAc:MeOH 90:10) was carried out by prep. TLC on silica gel, using CHCl$_3$:MeOH:AcOH (9.0:1.0:0.1) and afforded syringaresinol-4-O-β-D-glucopyranoside (0.9 mg) (17) and pinoresinol-4-O-β-D-glucopyranoside (1.2 mg) (18). Combined groups HM to HP (HM'; 86.7 mg) were fractionated by RP$_{18}$-HPLC (MeOH 58%:H$_2$O 42%; flow rate: 1 mL/min) and yielded rosmarinic acid (Rt = 10.6 min, 2.2 mg) (12) and p-anisic acid (Rt = 18.9 min, 0.1 mg) (6), and the sub-fraction HM'14b (Rt = 16.7 min, 5.9 mg), which was further purified by prep. TLC on silica gel with EtOAc:AcOH:H$_2$O (65:15:20) and led to the isolation of hesperidin (3.0 mg) (16) and of 4-hydroxyacetophenone 4-O-(3,5-dimethoxy-4-hydroxybenzoyl)-β-D-glucopyranoside (2.8 mg) (11). Combined groups HT to HV (HT'; 176.5 mg), subjected to CC on cellulose (11.0 cm × 3.2 cm) using as eluent AcOH:H$_2$O (30:70) afforded 79 fractions. Fraction HT'H (4.8 mg) was purified by prep. TLC on silica gel with EtOAc:AcOH:H$_2$O (65:15:20) to obtain 6-hydroxyluteolin 7-O-β-D-glucoside (0.9 mg) (15). The purification of fraction N (248.5 mg) was performed on silica gel CC (15.2 cm × 2.0 cm) using mixtures of cyclohexane: DM:EtOAc:MeOH of increasing polarity and afforded methyl α-linolenate (C18:3) (1.4 mg) (20) and ursolic acid (4.7 mg) (1). Groups NB' (7.4 mg) and NF (19.6 mg) were subjected to prep. TLC on silica gel with CHCl$_3$:MeOH:AcOH (9.5:0.5:0.05) and yielded linoleic acid (C18:2) (0.5 mg) (19) and maslinic acid (18.3 mg) (2), respectively. All obtained extracts, fractions, and isolated compounds were evaporated to dryness in vacuum under low temperature and then were put in activated desiccators with P$_2$O$_5$ until their weights had stabilized.

4.4. NMR Data of 5, 8, and 13

Compound 5: Yellow amorphous powder; ^{1}H-NMR (CD$_3$OD, 400 MHz): 2.66 (2H, t, J = 7.3, H-7), 3.66 (2H, t, J = 7.3, H-8), 6.52 (1H, dd, J = 8.0, 2.0, H-6), 6.65 (1H, d, J = 2.0, H-2), 6.67 (1H, d, J = 8.0, H-5).

Compound 6: White amorphous powder; ^{1}H-NMR (CD$_3$OD, 400 MHz): 3.91 (3H, s, OCH$_3$), 6.85 (2H, d, J = 8.5, H-3/H-5), 7.89 (2H, d, J = 8.5, H-2/H-6).

Compound 8: White amorphous powder; ^{1}H-NMR (CD$_3$OD, 400 MHz): δ 3.38 (1H, m, H-5'), 3.47 (2H, m, H-2', H-3'), 3.56 (2H, s, H-5a,b''), 3.61 (1H, m, H-6b'), 3.73 (1H, d, J = 9.6, H-4b''), 3.90 (1H, d, J = 2.2, H-2''), 3.96 (1H, d, J = 9.6, H-4a''), 3.99 (1H, d, J = 11.5, H-6a'), 4.96 (1H, d, J = 2.2, H-1''), 5.64 (1H, d, J = 7.9, H-1'), 6.84 (2H, d, J = 8.7, H-3, H-5), 7.96 (2H, d, J = 8.7, H-2, H-6).

Compound 13: Yellow amorphous powder; $[α]_D^{20}$ + 0.36° (c 0.350 MeOH); ^{1}H-NMR (CD$_3$OD, 400 MHz): δ 2.90 (1H, dd, J = 14.0, 9.8, H-7'), 3.07 (1H, dd, J = 14.0, 3.8, H-7'), 4.24 (1H, d, J = 6.5, H-8''), 4.88 (1H, d, J = 6.5, H-7''), 5.02 (1H, dd, J = 9.8, 3.8, H-8'), 6.31 (1H, d, J = 16.0, H-8), 6.38 (1H, d, J = 8.0, H-5), 6.62 (1H, dd, J = 8.0, 1.6, H-6'), 6.66 (1H, d, J = 8.0, H-5'), 6.74 (1H, d, J = 8.0, H-5''), 6.76 (1H, s, H-2'), 6.82 (1H, dd, J = 8.0, 2.0, H-6), 6.84 (1H, dd, J = 8.0, 2.0, H-6''), 6.98 (1H, d, J = 2.0, H-2''), 7.01 (1H, d, J = 2.0, H-2), 7.47 (1H, d, J = 16.0, H-7).

Supplementary Materials: The following are available online at www.mdpi.com/2223-7747/7/1/18/s1. Table S1: Non-volatile secondary metabolites of *Salvia* L.

Acknowledgments: The authors thank J. Heilmann (University of Regensburg, Pharmaceutical Biology), as well as J. Kiermaier for recording the MS spectra (all central analytics of NWF IV, University of Regensburg, Germany).

Author Contributions: T.M. contributed to the writing and carried out all chemical analyses. H.S. conceived and designed the experiments, contributed to the writing, and also supervised the chemical analyses.

Conflicts of Interest: The authors declare no conflict of interest.

References

1. Lu, Y.; Foo, L.Y. Polyphenolics of *Salvia*—A review. *Phytochemistry* **2002**, *59*, 117–140. [CrossRef]
2. Walker, J.B.; Systma, K.J. Staminal Evolution in the genus *Salvia* (Lamiaceae). *Ann. Bot.* **2007**, *100*, 375–391. [CrossRef] [PubMed]
3. Wu, Y.-B.; Ni, Z.-Y.; Shi, Q.-W.; Dong, M.; Kiyota, H.; Gu, Y.-C.; Cong, B. Constituents from *Salvia* Species and Their Biological Activities. *Chem. Rev.* **2012**, *112*, 5967–6026. [CrossRef] [PubMed]
4. Kamatou, G.P.P.; Makunga, N.P.; Ramogola, W.P.N.; Viljoena, A.M. South African *Salvia* species: A review of biological activities and phytochemistry. *J. Ethnopharmacol.* **2008**, *119*, 664–672. [CrossRef] [PubMed]
5. Grieve, M. *A Modern Herbal*; Savas Publishing: El Dorado Hills, CA, USA, 1984; ISBN 13 9780486227986.
6. Dereboylu, A.; Şengonca, N.; Güvensen, A.; Gücel, S. Anatomical and palynological characteristics of *Salvia willeana* (Holmboe) Hedge and *Salvia veneris* Hedge endemic to Cyprus. *Afr. J. Biotechnol.* **2010**, *9*, 2076–2088.
7. Bellomaria, B.; Arnold, N.; Valentini, G. Contribution to the study of the essential oils from three species of *Salvia* growing wild in the eastern Mediterranean region. *J. Essent. Oil Res.* **1992**, *4*, 607–614. [CrossRef]
8. Vonaparti, A.; Karioti, A.; Recio, M.; Máñez, S.; Ríos, J.; Skaltsa, E.; Giner, R. Effects of terpenoids from *Salvia willeana* in delayed-type hypersensitivity, human lymphocyte proliferation and cytokine production. *Nat. Prod. Commun.* **2008**, *3*, 1953–1958.
9. De la Torre, M.; Bruno, M.; Savona, F.; Rodriquez, B.; Apostolides Arnold, N. Terpenoids from *Salvia willeana* and *S. virgata*. *Phytochemistry* **1990**, *29*, 668–670. [CrossRef]
10. Wang, M.; Kikuzaki, H.; Lin, C.; Kahyaoglu, A.; Huang, M.; Nakatani, N.; Ho, C. Acetophenone Glycosides from Thyme (*Thymus vulgaris* L.). *J. Agric. Food Chem.* **1999**, *47*, 1911–1914. [CrossRef] [PubMed]
11. Machida, K.; Yogiashi, Y.; Matsuda, S.; Suzuki, A.; Kikuchi, M. A new phenolic glycoside syringate from the bark of *Juglans mandshurica* MAXIM. var. *sieboldiana* MAKINO. *J. Nat. Med.* **2009**, *63*, 220–222. [CrossRef] [PubMed]
12. Wang, M.; Kikuzaki, H.; Zhu, N.; Sang, S.; Nakatani, N.; Ho, C.-T. Isolation and structural elucidation of two new glycosides from sage (*Salvia officinalis* L.). *J. Agric. Food Chem.* **2000**, *48*, 235–238. [CrossRef] [PubMed]
13. Olennikov, D.; Chekhirova, G.V. 6″-Galloylpicein and other phenolic compounds from *Arctostaphylos uva-ursi*. *Chem. Nat. Compd.* **2013**, *49*, 1–7. [CrossRef]
14. Özgen, U.; Sevindik, H.; Kazaz, C.; Yigit, D.; Kandemir, A.; Secen, H.; Calis, I. A new sulfated a-Ionone glycoside from *Sonchus erzincanicus* Matthews. *Molecules* **2010**, *15*, 2593–2599. [CrossRef] [PubMed]
15. Benayache, F.; Boureghda, A.; Ameddah, S.; Marchioni, E.; Benayache, S. Flavonoids from *Thymus numidicus* Poiret. *Pharm. Lett.* **2014**, *6*, 50–54.
16. Rashed, K.; Ćirić, A.; Glamočlija, J.; Calhelha, R. Antimicrobial and cytotoxic activities of *Alnus rugosa* L. aerial parts and identification of the bioactive components. *Ind. Crops Prod.* **2014**, *59*, 189–196. [CrossRef]
17. Lu, Y.; Foo, L.Y. Flavonoid and phenolic glycosides from *Salvia officinalis*. *Phytochemistry* **2000**, *55*, 263–267. [CrossRef]
18. Ikan, R. *Natural Products. A Laboratory Guide*, 2nd ed.; Academic Press Inc.: Cambridge, MA, USA, 1991; pp. 9–11.
19. Carvalho, M.G.; Costa, P.M.; Santos Abreu, H. Flavones from *Vernonia diffusa*. *J. Braz. Chem. Soc.* **1999**, *10*, 163–166. [CrossRef]
20. Marder, M.; Viola, H.; Wasowski, C.; Fernández, S.; Medina, J.; Paladini, A. 6-Methylapigenin and hesperidin, new *valeriana* flavonoids with activity on the CNS. *Pharmacol. Biochem. Behav.* **2003**, *75*, 537–545. [CrossRef]
21. Aghel, N.; Ramezani, Z.; Beiranvard, S. Hesperidin from *Citrus sinensis* Cultivated in Dezful, Iran. *Pak. J. Biol. Sci.* **2008**, *11*, 2451–2453. [CrossRef] [PubMed]
22. Nizamutdinova, I.T.; Jeong, J.J.; Xu, G.H.; Lee, S.; Kang, S.S.; Kim, Y.S.; Chang, K.C.; Kim, H.J. Hesperidin, hesperidin methyl chalone and phellopterin from *Poncirus trifoliata* (Rutaceae) differentially regulate the expression of adhesion molecules in tumor necrosis factor-α-stimulated human umbilical vein endothelial cells. *Int. Immunopharmacol.* **2008**, *8*, 670–678. [CrossRef] [PubMed]
23. Maltese, F.; Erkelens, C.; Kooy, F.; Choi, Y.; Verpoorte, R. Identification of natural epimeric flavanone glycosides by NMR spectroscopy. *Food Chem.* **2009**, *116*, 575–579. [CrossRef]
24. Han, S.; Mok, S.; Kim, H.; Lee, J.; Lee, D.; Lee, S.Y.; Kim, J.; Kim, S.; Lee, S. Determination of hesperidin in mixed tea by HPLC. *J Agric. Sci.* **2011**, *38*, 295–299.
25. Lee, C.K.; Chang, M.H. The chemical constituents from the heartwood of *Eucalyptus citriodora*. *J Chin. Chem. Soc.* **2000**, *47*, 555–560. [CrossRef]

26. Hatzakis, E.; Agiomyrgianaki, A.; Kostidis, S.; Dais, P. High-resolution NMR spectroscopy, an alternative fast tool for Qualitative and Quantitative Analysis of Diacylglycerol (DAG) oil. *J. Am. Oil Chem. Soc.* **2011**, *88*, 1695–1708. [CrossRef]
27. Horst, M.; Urbin, S.; Burton, R.; MacMillan, C. Using proton nuclear magnetic resonance as a rapid response research tool for methyl ester characterization in biodiesel. *Lipid Technol.* **2009**, *21*, 1–3. [CrossRef]
28. Güvenalp, Z.; Özbek, H.; Kuruüzüm-Uz, A.; Kazaz, C.; Demirezer, Ö.L. Secondary metabolites from *Nepeta heliotropifolia*. *Turk. J. Chem.* **2009**, *33*, 667–675.
29. Ragasa, Y.; Ng, V.A.S.; Ebajo, V.D.; Fortin, D.R.; De Los Reyes, M.M.; Shen, C.-C. Triterpenes from *Shorea negrosensis*. *Pharm. Lett.* **2014**, *6*, 453–458.
30. Raza, R.; Ilyas, Z.; Ali, S.; Muhammad, N.M.; Muhammad, Y.; Khokhar, M.Y.; Iqbal, J. Identification of Highly Potent and Selective α-Glucosidase Inhibitors with Antiglycation Potential, Isolated from *Rhododendron arboreum*. *Rec. Nat. Prod.* **2015**, *9*, 262–266.
31. Tanaka, J.C.A.; Vidotti, G.J.; da Silva, C.C. A New Tormentic Acid Derivative from *Luehea divaricata* Mart. (Tiliaceae). *J. Braz. Chem. Soc.* **2003**, *14*, 475–478. [CrossRef]
32. Rungsimakan, S. Phytochemical and Biological Activity Studies on *Salvia viridis* L. Ph.D. Thesis, Department of Pharmacy and Pharmacology, University of Bath, Bath, UK, 2011.
33. Woo, W.K.; Han, Y.J.; Un Choi, S.; Ki, H.; Kim, K.H.; Lee, K.R. Triterpenes from *Perilla frutescens* var. *acuta* and Their Cytotoxic Activity. *Nat. Prod. Sci.* **2014**, *20*, 71–75.
34. Manns, D. Linalool and cineole type glucosides from *Cunila spicata*. *Phytochemistry* **1995**, *39*, 1115–1118. [CrossRef]
35. Wang, M.; Shao, Y.; Huang, T.; Wei, G.; Ho, C. Isolation and structural elucidation of aroma constituents bound as glucosides from Sage (*Salvia officinalis*). *J. Agric. Food Chem.* **1998**, *46*, 2509–2511. [CrossRef]
36. Pabst, A.; Barron, D.; Sémon, E.; Schreier, P. Two diastereomeric 3-oxo-α-ionol β-D-glucosides from Raspberry fruit. *Phytochemistry* **1992**, *31*, 1649–1652. [CrossRef]
37. Takeda, Y.; Zhang, H.; Matsumoto, T.; Otsuka, H.; Oosio, Y.; Honda, G.; Tabata, M.; Fujita, T.; Su, H.; Sezik, E.; et al. Megastigmane glycosides from *Salvia nemorosa*. *Phytochemistry* **1997**, *44*, 117–120. [CrossRef]
38. Sugiyama, M.; Nagayama, E.; Kikuchi, M. Lignan and Phenylpropanoid glycosides from *Osmanthus asiaticus*. *Phytochemistry* **1993**, *33*, 1215–1219. [CrossRef]
39. Kim, M.; Moon, H.; Lee, D.; Woo, E. A new lignan glycoside from the Stem bark of *Styrax japonica* S. et Z. *Arch. Pharm. Res.* **2007**, *30*, 425–430. [CrossRef] [PubMed]
40. Qin, Y.; Yin, C.; Cheng, Z. A new tetrahydrofuran lignan diglycoside from *Viola tianshanica* Maxim. *Molecules* **2013**, *18*, 13636–13644. [CrossRef] [PubMed]
41. Lee, J.; Lee, M.; Pae, S.; Oh, K.; Jung, C.; Baek, I.; Lee, S. Analysis of yield of eleutherosides B and E in *Acanthopanax divaricatus* and *A. koreanum* Grown with varying cultivation methods. *Sci. World J.* **2014**. [CrossRef]
42. Lall, N.; Kishore, N.; Binneman, B.; Twilley, D.; Venter, M.; Plessis-Stoman, D.; Boukes, G.; Hussein, A. Cytotoxicity of Syringin and 4-methoxycinnamyl alcohol isolated from *Foeniculum vulgare* on selected human cell lines. *Nat. Prod. Res.* **2015**, 1752–1756. [CrossRef] [PubMed]
43. Mulkens, A.; Kapetanidis, I. Eugenylglucoside, a new natural phenylpropanoid heteroside from *Melissa officinalis*. *J. Nat. Prod.* **1988**, *51*, 496–498. [CrossRef] [PubMed]
44. Ushiyama, M.; Furuya, T. Glycosylation of phenolic compounds by root culture of *Panax ginseng*. *Phytochemistry* **1989**, *28*, 3009–3013. [CrossRef]
45. Lu, Y.; Foo, L.Y. Rosmarinic acid derivatives from *Salvia officinalis*. *Phytochemistry* **1999**, *51*, 91–94. [CrossRef]
46. Woo, E.; Piao, M. Antioxidative Constituents from *Lycopus lucidus*. *Arch. Pharm. Res.* **2004**, *27*, 173–176. [CrossRef] [PubMed]
47. Özgen, U.; Mavi, A.; Terzi, Z.; Kazaz, C.; Aşçi, A.; Kaya, Y.; Seçen, H. Relationship between chemical structure and antioxidant activity of luteolin and its glyxosides isolated from *Thymus sipyleus* subsp. *sipyleus* var. *sipyleus*. *Rec. Nat. Prod.* **2011**, *5*, 12–21.
48. Dapkevicius, A.; Beek, T.; Lelyveld, G.; Veldhuizen, A.; Groot, A.; Linssen, J.; Venskutonis, R. Isolation and structure Elucidation of radican scavengers from *Thymus vulgaris* Leaves. *J. Nat. Prod.* **2002**, *65*, 892–896. [CrossRef] [PubMed]
49. Shahat, A.A.; Abdel-Azim, N.S.; Pieters, L.; Vlietinck, A.J. Isolation and NMR spectra of syringaresinol-β-D-glucoside from *Cressa cretica*. *Fitoterapia* **2004**, *75*, 771–773. [CrossRef] [PubMed]

50. Park, H.B.; Lee, K.H.; Kim, K.H.; Lee, I.K.; Noh, H.J.; Choi, S.U.; Lee, K.R. Lignans from the roots of *Berberis amurensis*. *Nat. Prod. Sci.* **2009**, *15*, 17–21.
51. Klimek, B.; Tokar, M. Biologically active compounds from the flowers of *Forsythia suspensa* Vahl. *Acta Pol. Pharm.* **1998**, *55*, 499–504.
52. Ouyang, M.; Shung, Y.; Zhang, Z.K.; Kuo, Y.H. Inhibitory Activity against Tobacco Mosaic Virus (TMV) Replication of Pinoresinol and Syringaresinol Lignans and Their Glycosides from the Root of *Rhus javanica* var. *roxburghiana*. *J. Agric. Food Chem.* **2007**, *55*, 6460–6465. [CrossRef] [PubMed]
53. Chiou, W.; Shen, C.; Lin, L. Anti-inflammatory principles from *Balanopora laxiflora*. *J. Food Drug Anal.* **2011**, *19*, 502–508.
54. Kim, A.R.; Ko, H.J.; Chowdhury, M.A.; Chang, Y.; Woo, E. Chemical constituents of the aerial parts of *Artemisia selengensis* and their IL-6 inhibitory activity. *Arch. Pharm. Res.* **2015**, *38*, 1059–1065. [CrossRef] [PubMed]
55. Jerković, I.; Mastelić, J.; Marijanović, Z. A variety of volatile compounds as Markers in Unifloral Honey from Dalmatian Sage (*Salvia officinalis* L.). *Chem. Biodivers.* **2006**, *3*, 1307–1316. [CrossRef] [PubMed]
56. Wu, S.; Chan, Y. Five new iridoids from roots of *Salvia digitaloides*. *Molecules* **2014**, *19*, 15521–15534. [CrossRef] [PubMed]
57. Mabry, T.G.; Markham, K.R.; Thomas, M.B. *The Systematic Identification of Flavonoids*; Springer: New York, NY, USA, 1970.
58. Coll, J.C.; Bowden, B.F. The application of Vacuum Liquid Chromatography to the separation of terpene mixtures. *J. Nat. Prod.* **1986**, *49*, 934–936. [CrossRef]
59. Stahl, E. *Thin-Layer Chromatography, a Laboratory Handbook*, 2nd ed.; Springer: Berlin/Heidelberg, Germany, 1969.
60. Neu, R. Chelate von Diarylborsäuren mit aliphatischen Oxyalkylaminen als Reagenzien für den Nachweis von Oxyphenyl-benzo-γ-pyronen. *Naturwissenschaften* **1957**, *44*, 181–183. [CrossRef]

 © 2018 by the authors. Licensee MDPI, Basel, Switzerland. This article is an open access article distributed under the terms and conditions of the Creative Commons Attribution (CC BY) license (http://creativecommons.org/licenses/by/4.0/).

Article

Green Ultrasound Assisted Extraction of *trans* Rosmarinic Acid from *Plectranthus scutellarioides* (L.) R.Br. Leaves

Duangjai Tungmunnithum [1,2,3,*], Laurine Garros [1,2,4], Samantha Drouet [1,2], Sullivan Renouard [5], Eric Lainé [1,2] and Christophe Hano [1,2,*]

1. Laboratoire de Biologie des Ligneux et des Grandes Cultures, INRA USC1328, Orleans University, 45067 Orléans Cedex 2, France; laurine.garros@etu.univ-orleans.fr (L.G.); samantha.drouet@univ-orleans.fr (S.D.); eric.laine@univ-orleans.fr (E.L.)
2. Bioactifs et Cosmetiques, CNRS GDR 3711 Orleans, 45067 Orléans Cedex 2, France
3. Department of Pharmaceutical Botany, Faculty of Pharmacy, Mahidol University, Bangkok 10400, Thailand
4. Institut de Chimie Organique et Analytique, CNRS UMR731, Orleans University, 45067 Orléans Cedex 2, France
5. Institut de Chimie et de Biologie des Membranes et des Nano-objets, CNRS UMR 5248, Bordeaux University, 33600 Pessac, France; sullivan.renouard@u-bordeaux.fr
* Correspondence: duangjai.tun@mahidol.ac.th (D.T.); hano@univ-orleans.fr (C.H.); Tel.: +662-6448677-91 (D.T.); Tel.: +33-237-309-753 (C.H.)

Received: 30 December 2018; Accepted: 23 February 2019; Published: 27 February 2019

Abstract: Painted nettle (*Plectranthus scutellarioides* (L.) R.Br.) is an ornamental plant belonging to *Lamiaceae* family, native of Asia. Its leaves constitute one of the richest sources of *trans*-rosmarinic acid, a well-known antioxidant and antimicrobial phenolic compound. These biological activities attract interest from the cosmetic industry and the demand for the development of green sustainable extraction processes. Here, we report on the optimization and validation of an ultrasound-assisted extraction (USAE) method using ethanol as solvent. Following preliminary single factor experiments, the identified limiting extraction parameters (i.e., ultrasound frequency, extraction duration, and ethanol concentration) were further optimized using a full factorial design approach. The method was then validated following the recommendations of the association of analytical communities (AOAC) to ensure the precision and accuracy of the method used to quantify *trans*-rosmarinic acid. Highest *trans*-rosmarinic acid content was obtained using pure ethanol as extraction solvent following a 45-minute extraction in an ultrasound bath operating at an ultrasound frequency of 30 kHz. The antioxidant (in vitro radical scavenging activity) and antimicrobial (directed toward *Staphylococcus aureus* ACTT6538) activities were significantly correlated with the *trans*-rosmarinic acid concentration of the extract evidencing that these key biological activities were retained following the extraction using this validated method. Under these conditions, 110.8 mg/g DW of *trans*-rosmarinic acid were extracted from lyophilized *P. scutellarioides* leaves as starting material evidencing the great potential of this renewable material for cosmetic applications. Comparison to other classical extraction methods evidenced a clear benefit of the present USAE method both in terms of yield and extraction duration.

Keywords: *Plectranthus scutellarioides*; *trans*-rosmarinic acid; *Lamiaceae*; green extraction; ultrasound; antioxidant; antimicrobial

1. Introduction

Plectranthus scutellarioides (L.) R.Br. or painted nettle (Figure 1a) is an ornamental and medicinal plant, mainly found in Asia, currently spread over America, Southern Africa, and Europe [1,2].

Its synonym, *Coleus blumei* Benth. is also well-known. This plant is a species member of the *Lamiaceae* family, widely cultivated as an ornamental plant due to the beauty of its leaves. Furthermore, local people in many countries have long been used *P. scutellarioides* as an herbal medicine. For example, local Indian people in Arunachal Pradesh use hot water extract of *P. scutellarioides* mixing with fruit juice to apply on the skin in order to decrease undesirable symptoms after scorpion bite [3]. In addition, Tlanchinol people in Mexico use the infusion of this plant for gastrointestinal diseases treatment [4]. Interestingly, its leaves constitute one of the richest sources of an antioxidant and antimicrobial phenolic compound, namely *trans*-rosmarinic acid (RA) (Figure 1b) [5–8].

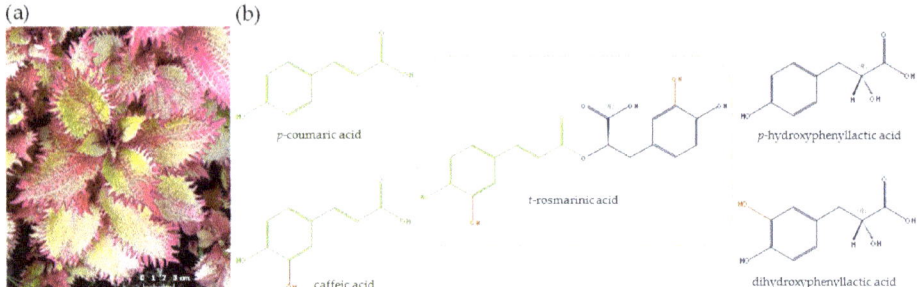

Figure 1. (**a**) Painted nettle (*Plectranthus scutellarioides* (L.) R.Br.) leaves morphology; (**b**) Chemical structures of *p*-coumaric acid, caffeic acid, *p*-hydrophenyllactic acid, dihydrophenyllactic acid, and *trans*-rosmarinic acid (picture taken by C.H.).

From a strict chemical point of view, RA is a caffeic acid and dihydrophenyllactic acid ester; however, biologically, its biosynthetic precursors are *p*-coumaric acid and *p*-hydroxyphenyllactic acid, respectively (Figure 1b). RA has been identified as one of the main bioactive compounds in many *Lamiaceae* from the Nepetoideae subfamily [8], and beside *P. scutellarioides* other *Plectranthus* species have been described for their accumulation of RA such as *P. barbatus* [9–12], *P. vericallatus* [9,12], *P. madagascarensis* [12,13], *P. ecklonii* [9], *P. fructicosus* [9], *P. lanuginosus* [9], *P. hadiensis* [12], *P. neochilus* [12], *P. amboinicus* [14], or *P. ornatus* [15]. These studies focusing on the biological activities associated with the presence of RA in the extracts obtained from these *Plectranthus* species or comparison of RA content variation among some of these species. Indeed, a number of relevant biological activities have been ascribed to this natural compound among which antioxidant [16], antimicrobial [7], anti-inflammatory [17], antimutagenic [8], antiagiogenic [8], neuroprotective [18], Alzheimer disease preventive action [19] with acetylcholinesterase inhibition capacity [10,11].

Antioxidant action is of particular interest since excessive accumulation of free radicals could constitute a starting point or aggravating factor for many diseases though their potential damages on membrane lipids, DNA, and proteins. Today, natural antioxidants are considered as potential safer and efficient drugs to prevent a wide range of diseases resulting from oxidative stress [20]. Indeed, damaging effects on health, including carcinogenesis, of their synthetic antioxidant and preservative counterparts have been pointed out [21]. As a consequence, the uses of some of these synthetic compounds are now strictly regulated, some have been removed from the 'generally recognized as safe' (GRAS) list and are now forbidden for food applications in Japan, Canada, and Europe [22]. Natural compounds have therefore attracted attention because of their potential for application to the food, cosmetic and pharmaceutical industries as natural preservatives because of their antioxidant and antibacterial activities [23,24]. However, the development of effective extraction methods of these natural compounds is necessary. Many extraction methods have been developed to extract natural antioxidants from various naturally occurring matrices based on maceration extraction, Soxhlet extraction, microwave assisted extraction, or ultrasound-assisted extraction (USAE) [25–27]. Green extraction technologies have attracted high interest in modern industries over the last decade and ultrasound-assisted extraction (USAE) is now considered as one of the most efficient energy-saving process in terms of duration, selectivity, and reproducibility, operating under soft- to mid-extraction conditions [25]. The improvement of extraction efficiency obtained using USAE is reported to rely on both acoustic cavitation and mechanical effects [25]. Indeed, ultrasounds (US) produce an acoustic cavitation effect facilitating the penetration of the extraction solvent. Consequently, an easier release of the intracellular content of the plant material is observed thank to a greater solvent agitation resulting in an increased surface contact between the solvent and the target compound as well as an enhanced solubility of the target compound into the extraction solvent [25].

To date, most of the studies dealing with RA production from *P. scutellarioides* have focused on the elucidation of its biosynthetic pathway [8] and/or biotechnological approaches to improve its production in planta using in vitro technologies [5,7,8,28–30]. Little attention has been paid to the optimization of its extraction from the leaves of this ornamental plant. This ornamental plant is known as easy to propagate by cuttings and high renewable biomass production of leaves can be obtained through basic horticultural approaches, therefore the development of green extraction of RA using this raw starting material for potential cosmetic applications could be very promising. In the literature, some studies have reported on the optimization of USAE of RA from the leaves of other *Lamiaceae* plants such as *Hyssopus cuspidatus* [31], *Rosmarinus officinalis* [32,33], *Melissa officinalis* [34], *Mentha piperita*, *Mentha longifolia*, or *Ocimum basilicum* [35]. RA contents reported by these studies evidenced a strong matrix effect. Plant matrix can have considerable effects (so-called matrix effects) both on the optimal extraction conditions and the resulting extraction yields [36]. This matrix effect depends on the plant species, plant origins, organs or tissue used, drying and storage conditions, and so on. For example, RA obtained following USAE of *Rosmarinus officinalis* dried leaves greatly varied according to the extraction conditions used, as determined by three independent studies [32,33,37], thus evidencing the necessity to specifically optimize and validate the extraction conditions for each plant matrix. In the present study, our goals were to develop and validate a green USAE protocol of RA from *P. scutellarioides* leaves for future cosmetic applications as natural antioxidant and natural preservative using ethanol as green extraction solvent. For this purpose, in order to extract high amount of RA from *P. scutellarioides* leaves we have determined of the most critical extraction parameters—among ethanol concentration, US frequencies, solvent to material ratio, extraction temperature and extraction duration, applied a factorial design of experiment taking into account the possible interaction between the limiting extraction parameters, determined the optimal extraction conditions after statistical analysis and using 3D surface plots, and validated the optimal extraction conditions. Then antioxidant and antibacterial activities of each extract were determined using, the DPPH assay and a validated 96-well plate assay for the monitoring of *Staphylococcus aureus* ACTT6538 growth inhibition, respectively. The efficiency of this optimized and validated extraction protocol was finally compared to the conventional heat reflux extraction procedures.

2. Results and Discussion

2.1. Preliminary Single Factor Experiments

The relative effect of different independent parameters on the extraction yield of RA from *P. scutellarioides* leaves was first evaluated through single-factor experiments. Here, the effects of several parameters described as important in the literature have been studied: ethanol concentration in aqueous solution, extraction time, extraction temperature, ultrasound frequency, and solvent-to-material ratio (S/M ratio). The main objective of these preliminary single factor experiments was to determine the main limiting factors that will be then applied in the design of experiment in order to evaluate the interaction effect of these extraction parameters.

As part of a green chemistry approach development, the choice of extraction solvent is one of the crucial parameters to take into account. Various organic solvents have commonly been used to extract antioxidant polyphenols from a various plant matrix, such as methanol, ethanol, and acetone [26].

Among these solvents, ethanol is non-toxic to humans and environmentally friendly. Its extraction efficiency can further be improved by mixing it with water, thus making it able to extract a wide range of phenolic compounds. These two universal solvents also present the great advantage of being inexpensive and are therefore widely used in the agri-food and/or cosmetic industries [25–27]. For all these reasons, we chose ethanol as the extraction solvent and tested the influence of ethanol concentration in aqueous solution on the extraction of RA. Preliminary experiments were conducted with various concentrations of aqueous ethanol solutions (0, 25, 50, 75, and 100% (v/v)) using fixed S/M ratio (25:1 mL/g DW), extraction time (30 min), sonication frequency (30 kHz), and extraction temperature (45°C). The results shown in Figure 2a indicate that the RA extraction yield increases with the ethanol concentration reaching a peak (82.1 ± 6.5 mg / g DW) with pure ethanol (100%). The reasons for this could be related to the solubility of RA, which LogP of 2.4 indicates higher solubility in octanol than in water, and to the polarity of ethanol.

It is well established that the ultrasound frequency could greatly influence the extraction efficiency through cavitation effect and by influencing diffusion coefficient of the targeted compounds, resulting in the solubility improvement of the target compound in the extraction solvent. From this observation, it appears that the extraction efficiency could be improved by an increase of the ultrasonic frequency. Nevertheless, high ultrasound frequency can, on the contrary, lead in some conditions to the degradation of the bioactive compounds and consequently considerably reduce the extraction yield [25–27]. Therefore, the ultrasound frequency must be precisely optimized. The effect of different ultrasound frequency (0, 15, 30, and 45 kHz) on the extraction yield of RA was thus next evaluated using fixed parameters set as follows: ethanol concentration 100%, S/M ratio 25:1 mL/g DW, extraction time 45 min, and extraction temperature of 45 °C. According to the results presented in Figure 2b, application of ultrasound frequency of 30 and 45 kHz here improved the RA extraction.

Different solvent to material (S/M) ratios were then evaluated (10:1, 25:1, and 50:1 ml of pure ethanol per gram of DW material). These experiments were conducted using fixed ethanol concentration (100% (v/v)), extraction time (30 min), sonication frequency (30 kHz), and extraction temperature (45 °C). The results shown in Figure 2c indicate that the RA extraction yield was not significantly influenced by the S/M ratio.

The effect of different extraction temperature (25, 35, 45, 55, 65, and 75°C) on the extraction yield of RA was thus next evaluated using fixed parameters set as follows: ethanol concentration 100%, S/M ratio 25:1 mL/g DW, extraction time 45 min and ultrasound frequency 45 kHz. According to the results presented in Figure 2d, the extraction efficiency is here only slightly influenced by the temperature parameter. Indeed, only a slight increase in the extraction yield was noted by increasing the extraction temperature from 25 to 65°C, whereas increasing extraction temperature to 75 °C using these conditions resulted in a decreased RA yield. It is accepted that an excessive temperature coupled to ultrasound can cause the degradation of the target compound [25–27]. From these results, extraction temperature did not appear as a limiting parameter for the extraction of RA from *P. scutellarioides* leaves.

The effect of extraction duration on the RA extraction efficiency was studied for a duration ranging from 0 to 60 min, with other parameters fixed at 100% for ethanol concentration, 25:1 mL/g (DW) for S/M ratio, 45°C as the extraction temperature and 30 kHz ultrasound frequency. The results depicted in Figure 2e indicate that the RA extraction yield increases with the extraction time in a first phase reaching a plateau after 45 min. We noted that the observed increase between 45 and 60 min was not statistically significant. Under these conditions, the maximum extraction efficiency could therefore be obtained after 45 min. It has been described that prolonged ultrasound can lead to the degradation of antioxidant phenolic compounds [25–27]. Consequently, extraction time was set to 45 min for the following experiments.

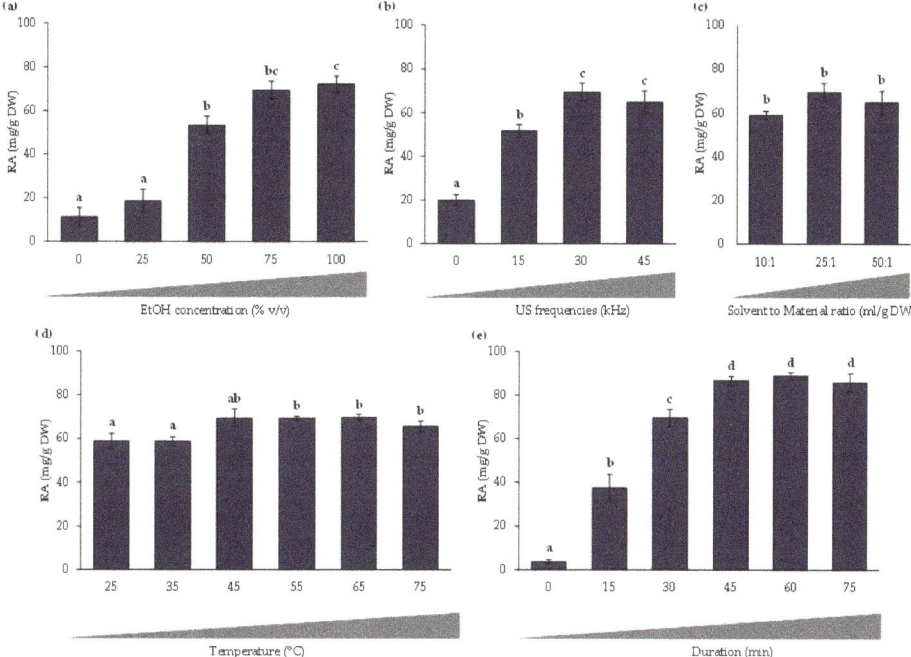

Figure 2. RA contents extracted from *P. scutellarioides* leaves as of function of (**a**) ethanol concentration, (**b**) ultrasound frequency, (**c**) solvent to material ratio, (**d**) extraction temperature, and (**e**) extraction duration. For a complete description of the extraction conditions, see text. Values are means ± SD of 3 independent replicates. Different letters (a–d) represent significant differences between the various extraction conditions ($p < 0.05$).

2.2. Develoment of a Multifactorial Approach

Experimental factorial design associated with statistical analysis and 3D surface response plots have proven their efficiencies in the precise and rapid optimization of extraction protocols by taking into account the possible interaction between independent variables which is not the case when developing an extraction protocol using a single factor approach [38]. Here, from our preliminary experiments, three influencing variables were selected to optimize RA extraction from *P. scutellarioides* leaves using a factorial design approach: aqueous ethanol concentration (X_1, ranging from 50 to 100 % (v/v)), US frequency (X_2, ranging from 15 to 45 kHz) and extraction duration (X_3, ranging from 15 to 45 min). The coded levels and actual experimental values of each independent variables are presented in Table 1. Note that, according to our preliminary experiments, a solvent to material ratio of 25:1 ml/g DW and an operating temperature of 45°C were chosen.

Table 1. Identities, code unit, coded levels and actual experimental values of the three independent variables.

Independent variable	Code unit	Coded variable levels		
		−1	0	+1
Ethanol concentration (% v/v) [1]	X_1	50	75	100
US frequency (kHz)	X_2	15	30	45
Extraction duration (min)	X_3	15	30	45

[1] % of ethanol concentration in mixture with HPLC grade ultrapure water.

Here, a full factorial design was chosen for the optimization because of the high reproducibility of the obtained results due to the real measurement of a extensive number of experimental conditions compared to other design of experiments approaches [39]. For the experiment, the 27 different conditions (run ID) using independent process variables were randomized (run order), tested as independent triplicates, and the resulting extracted RA content evaluated by HPLC (Table 2).

Table 2. Results of experimental design.

Run ID	Run order	X_1	X_2	X_3	RA (mg/g DW)
Obs1	17	−1	−1	−1	16.9 ± 2.6
Obs2	24	0	−1	−1	53.2 ± 0.5
Obs3	26	+1	−1	−1	57.8 ± 6.0
Obs4	21	−1	0	−1	18.7 ± 3.8
Obs5	22	0	0	−1	65.2 ± 5.0
Obs6	6	+1	0	−1	66.5 ± 2.4
Obs7	10	−1	+1	−1	17.6 ± 4.3
Obs8	27	0	+1	−1	78.2 ± 1.1
Obs9	7	+1	+1	−1	102.2 ± 1.15
Obs10	18	−1	−1	0	19.2 ± 2.2
Obs11	12	0	−1	0	55.1 ± 2.5
Obs12	8	+1	−1	0	72.6 ± 0.6
Obs13	25	−1	0	0	15.6 ± 0.1
Obs14	1	0	0	0	70.4 ± 7.4
Obs15	16	+1	0	0	91.6 ± 3.1
Obs16	23	−1	+1	0	20.3 ± 2.5
Obs17	11	0	+1	0	66.3 ± 2.6
Obs18	14	+1	+1	0	90.2 ± 1.3
Obs19	15	−1	−1	+1	16.9 ± 4.3
Obs20	3	0	−1	+1	56.7 ± 1.9
Obs21	13	+1	−1	+1	62.6 ± 1.6
Obs22	9	−1	0	+1	19.5 ± 0.5
Obs23	5	0	0	+1	91.1 ± 4.4
Obs24	19	+1	0	+1	110.8 ± 4.5 *
Obs25	4	−1	+1	+1	14.4 ± 1.1
Obs26	20	0	+1	+1	66.7 ± 3.0
Obs27	2	+1	+1	+1	75.9 ± 2.6

Values are the mean ± RSD of three independent replicates except for *, which correspond to the highest RA content here determined by six independent experiments to confirm this value.

Under these extraction conditions, we found RA contents extracted from freeze-dried *P. scutellarioides* leaves ranging from 2.4 (Obs22) to 110.8 (Obs24) mg/g DW. Following multiple regression analysis, the RA content (Y) as a function of the different variables (X_1, X_2, and X_3) was represented by the following second order polynomial equation: $Y = 73.22 + 31.74X_1 + 6.70X_2 + 2.15X_3 - 17.60X_1^2 - 8.68X_2^2 - 0.65X_3^2 + 6.34X_1X_2 + 2.11X_1X_3 - 4.11X_2X_3$ (Table 3).

Table 3. Statistical analysis of the regression coefficients.

Source	Value	SD	t	$P > |t|$
Constant	73.22	5.279	13.871	< 0.0001 ***
X_1	31.74	2.444	12.988	< 0.0001 ***
X_2	6.70	2.444	2.743	0.014 *
X_3	2.15	2.444	0.879	0.392 ns
X_1^2	−17.60	4.232	−4.159	0.001 **
X_2^2	−8.68	4.232	−2.052	0.046 *
X_3^2	−0.65	4.232	−0.153	0.880 ns
X_1X_2	6.34	2.993	2.118	0.049 *
X_1X_3	2.11	2.993	0.707	0.489 ns
X_2X_3	−4.11	2.993	−1.373	0.188 ns

SD standard error; *** significant $p < 0.001$; ** significant $p < 0.01$; * significant $p < 0.05$; ns not significant.

As a result of the statistical analysis, the linear coefficients X_1 and X_2, the quadratic coefficients X_1^2 and X_2^2 as well as the interaction coefficient X_1X_2 appeared to significantly influence the extraction efficiency of RA from freeze-dried *P. scutellarioides* leaves. On the contrary, the other linear (X_3), quadratic (X_3^2) and interaction (X_1X_3 and X_2X_3) coefficients were insignificant ($p > 0.05$). Therefore, ethanol concentration (X1) as well ultrasound frequency (X2) and their interaction appeared to influence greatly the extraction efficiency over extraction duration.

Analysis of variance (ANOVA) result and the fit of the obtained model are listed in Table 4. The model was highly significant as indicated by the high F-value (22.72) and the low *p*-value ($p < 0.0001$), but also by the low lack of fit F-value (0.87) and its non-significant associated *p*-value ($p > 0.05$). The RA contents predicted vs. the experimentally measured RA contents are plotted in Figure S1 and confirmed the high precision of the model. The determination coefficient R^2 of 0.924 and the adjusted R^2 of 0.883 both confirmed this model is adequate to predict the RA extraction yield from freeze-dried *P. scutellarioides* leaves. In addition, the variation coefficient value (CV = 0.79%) also indicated the adequacy between the model and experimental values.

Table 4. ANOVA of the predicted model for USAE of RA from freeze-dried *P. scutellarioides* leaves.

Source	Sum of square	df	Mean of square	F-value	*p*-value
Model	22074.5	9	2452.7	22.72	< 0.0001
Lack of fit	1592.8	17	93.7	0.87	0.266
Residual	1827.1	17	107.5	-	-
Pure Error	234.3	0	-	-	-
Cor. Error	23901.6	26	-	-	-
R^2	0.924				
R^2 adj	0.883				
CV %	0.79				

df: degree of freedom; Cor. Error: corrected error; R^2: determination coefficient; R^2 adj: adjusted R^2; CV variation coefficient value.

Here, the 3D plots accounted for the complexity of the USAE of RA from lyophilized *P. scutellarioides* leaves. All the linear coefficients (ethanol concentration, ultrasound frequency, and extraction duration) as well as the interaction coefficients X_1X_2 (ethanol concentration and ultrasound frequency) and X_1X_3 (ethanol concentration and extraction duration) of the second-order polynomial equation obtained were positive, indicating that increasing these parameters exerted a global favorable action on RA extraction. In good agreement with this observation, the resulting 3D plots confirmed the favorable positive effects on RA extraction resulting in the increase of ethanol concentration combined with the increased extraction duration or ultrasound frequency (Figure 3a,b). On the contrary, all the quadratic coefficients and the interaction coefficient X_2X_3 (ultrasound frequency and extraction duration) were negative. Thus, it clearly appeared that RA extraction according to these parameters passed by a maximum (Figure 3c). Here, prolonged extraction duration associated with high ultrasound frequency resulted in a decrease in RA content. This observed decrease could be due to partial degradation of RA. High ultrasound frequency is known to be potentially destructive and to induce oxidation of natural products that could lead to the loss of the biological activities of these compounds [25–27]. According to the adjusted second order polynomial equation, optimal conditions were 45 min extraction in an ultrasound bath operating at 36.8 kHz and using 99.8% (v/v) aqueous ethanol as extraction solvent. These conditions were adjusted to the material and highest RA content was therefore obtained following 45 minutes extraction in an ultrasound bath operating at 30 kHz and using pure ethanol as extraction solvent. Under these conditions, an RA content of 110.8 mg/g DW was measured from lyophilized *P. scutellarioides* leaves. The relative RA purity was estimated according to the method described by Falé et al. [11] for *P. barbatus* leaves decoction. Under our optimal extraction conditions, RA was estimated to represent around 70% of the total absorption of the wavelength using PDA detection. This was further confirmed by DEDL detection (data not shown). Our optimal conditions are quite similar to those described by Caleja et al. [34] for the USAE of RA from *R. officinalis* leaves, but comparatively our results showed a higher extraction yield with *P. scutellarioides* leaves as starting material. In the literature, RA extraction yields from *Lamiaceae* leaves using USAE greatly varied [40,41], ranging from 0.01 mg/g DW in *H. cuspidatus* [31] to 86.3 mg/g DW in *M. officinalis* [34], whereas a higher yield of 12.6 mg/g DW was obtained using *Marantha depressa* (*Maranthaceae*) leaves as starting material [42]. Comparatively, our results are in the high range of these values.

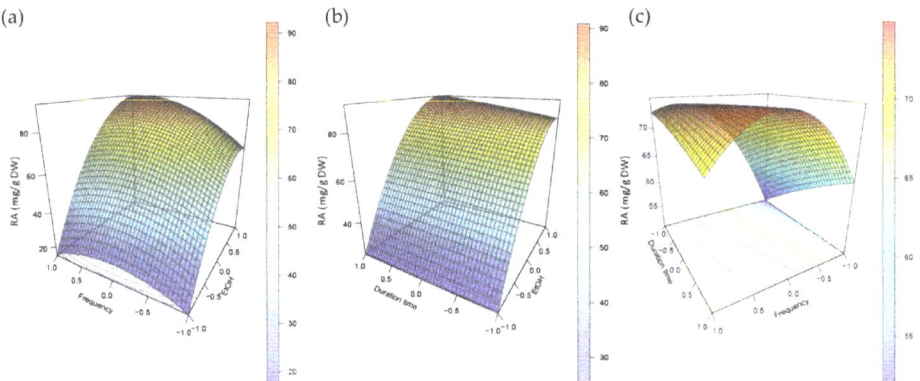

Figure 3. Predicted surface response plots of the RA extraction yield as a function of (**a**) ultrasound frequency and ethanol concentration, (**b**) extraction duration and ethanol concentration, and (**c**) extraction duration and ultrasound frequency.

2.3. Validation of the Extraction Method

Quantification of RA was achieved by HPLC analysis (Figure S2) and RA was identified by comparison with an authentic standard. A good separation resolution must be one of the primary goal for quantitative HPLC analysis. For this purpose, a separation resolution (R_S) of at least 1.5 should be obtained, with the exception of samples containing a high number of component, which is the case of raw plant extracts, or compounds more difficult to separate such as isomers [43]. Here, the presence of *cis*-RA, an isomer of *trans*-RA, resulted in a separation resolution of ca 1.3 (1.298). Considering that we here work with a raw plant extract and the nature of the 'interfering' peak, this resolution (higher than 1.0) is therefore acceptable according to international standards [43]. To ensure the accuracy and precision of the method used to quantify RA, the HPLC method was validated. The validation parameters are presented in Table 5.

Table 5. Validation parameters of the HPLC method.

Equation	R^2	LOD (ng)	LOQ (ng)	Precision (%RSD)		Repeatability (%RSD)	Recovery (%RSD)
				Intraday	Interday		
y = 4.872x − 0.123	0.9998	1.8	5.3	0.48	0.94	3.5	2.8

The six-point calibration curve of the peak areas (y) against the injected quantities of RA at 320 nm was linear over the wide range analyzed (50–1000 µg/ml) with $R^2 > 0.999$ and the slope of the standard covering the analytical range varied at most 1% relative standard deviation (RSD) over a four-week period. The LOD (S/N = 3) and the LOQ (S/N = 10) were as low as 1.8 ng and 5.3 ng. Then, the instrumental precision determination was obtained through five injections of the same sample. The method precision and stability were confirmed by the low RSD values of 0.48% and 0.94% obtained for interday and intraday precisions respectively (Table 5). The repeatabilty of the method was evidenced with a RSD value as low as 3.5%, as well as its accuracy, determined by four levels of h standard addition (from 5 to 50 µg/mL) showing a good recovery with a mean for the RSD values of 2.8% (Table 5).

2.4. Biological Activities of the Extracts

The next steps were the evaluation of the antioxidant and antimicrobial activities of the 27 extracts obtained using the full factorial design in order to confirm that the extraction did not alter the biological properties of RA and to establish correlations between these biological activities and the RA content. For this purpose, antioxidant activity was determined by the radical scavenging activity using the commonly used DPPH in vitro assay [44], whereas antimicrobial activity was determined by monitoring the growth inhibition of pathogenic strain of *Staphylococcus aureus* strain ACTT6538 using microplate method described by El Abdellaoui et al. [45]. The results are presented as a heatmap representation and classified using a hierarchical clustering analysis (HCA) using Euclidian distance (Figure 4). The individual values are presented in Table S1.

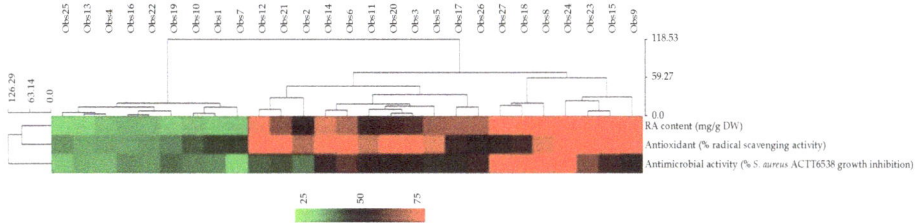

Figure 4. Hierarchical clustering analysis of RA contents and antioxidant (DPPH radical scavenging activity) and antimicrobial (growth inhibition of *S. aureus* ACTT6538) of the 27 extracts obtained following USAE of *P. scutellarioides* leaves.

Surface response plots showing the impact of the three independent parameters (ethanol concentration, ultrasound frequency, and extraction duration) are presented in Figure S3 for the radical scavenging activity and Figure S4 for the antimicrobial activity. These 3D plots depicted a very similar aspects to those obtained for RA extraction (Figure 3). Strong significant correlations were observed between RA content and antioxidant activity (Pearson coefficient correlation (PCC) = 0.836, $p = 0.004$ **) as well as between RA content and antimicrobial activity (PCC = 0.822, $p = 0.017$ *) were calculated confirming that the applied USAE conditions maintained RA into a bioactive form and preserved the associated biological activities. In particular the extract obtained using optimal conditions, containing the highest RA content, here offered the best radical scavenging activity (89.7% inhibition of radical formation) and antimicrobial action toward *S. areus* (79.8% growth inhibtion).

2.5. Comparison with Conventional Extraction Protocol and Other Biotechnological Sources

The last step of this work was to compare the optimal USAE of RA from *P. scutellarioides* leaves with a conventional heat reflux extraction method. For this purpose the same ethanol concentration (pure ethanol), temparature (45°C), and solid-to-liquid ratio (25:1) conditions were used, but using three different extraction duration of 45 min (i.e., the same extraction time used for the optimal USAE of RA), 90 min (i.e., 2-times the duration used for the optimal USAE of RA), and 180 min (i.e., 4-times the duration used for the optimal USAE of RA). The RA contents obtained are presented in Table 6.

Table 6. Comparison between conventional heat reflux method and ultrasound-assisted extraction (USAE) of RA from *P. scutellarioides* leaves.

RA content as a function of extraction method and time	Conventional Heat Reflux Extraction			USAE
Extraction duration	45 min	90 min	180 min	45 min
RA content (mg/g DW)	55.6 ± 3.1[d]	79.7 ± 1.5[c]	95.6 ± 2.2[b]	110.8 ± 4.5[a]

Values are the mean ± RSD of three independent replicates and different letters (a, b, c, d) indicate significant differences between conditions ($p < 0.05$).

The results of this comparison clearly indicated that USAE provided better RA extraction yield and in a shorter time compared with the conventional heat reflux method. Using the conventional heat reflux method, the RA content increased with extraction duration, but even after 180 minutes, the extraction yield is still lower than for USAE with a 3-times reduced extraction time. Note that for the same extraction duration the RA content obtained using the conventional heat reflux method is around half of this obtained using USAE. These results clearly evidenced that the use of USAE to extract RA from *P. scutellarioides* leaves is of great interest, in particular within the context of green chemistry, in term of reducing energy consumption by using this innovative technology. Here, USAE increased RA extraction yield and therefore lowered extraction costs (due to the reduction of treatment time and solvent consumption) certainly as a consequence of solvent heating by directly interacting with the water molecules present in the plant tissue resulting in a more efficient rupture of these tissue and release of RA into the solvent as it was hypothesized for other plants [25–27].

Many studies reported on RA production from *P. scutellarioides* using different biotechnological systems to improve its production from *in planta* using in vitro technologies, but little attention has been paid to the optimization of its extraction directly from the leaves of this ornamental plant. Here, using our optimized and validated USAE method, we report on RA yield of 11.08% DW from the leaves as starting material which is in the range of RA production obtained using biotechnological systems. Indeed, several authors reported on RA contents ranging from 2.85% (DW) following permeabilization using DMSO [30] to 19.00% (DW) after 4% sucrose feeding [29] using cell suspensions, or of 7.60% (DW) using hairy roots subjected to methyl jasmonate elicitation [28]. Considering *P. scutellarioides* as an ornamental plant that could easily be multiplied by cuttings, and its leaves are a cheap and renewable material, the present USAE method of RA, using *P. scutellarioides* leaves as raw starting material, is therefore of great interest for feasible industrial applications.

3. Materials and Methods

3.1. Chemicals and Reagents

Analytical grade extraction solvents were used in the present study and were obtained from Fisher Scientific (Illkirch, France). RA standard was purchased from Sigma-Aldrich (Saint-Quentin Fallavier, France). All chemicals for the purpose of antioxidant and antimicrobial activities were from Sigma-Aldrich (Saint-Quentin Fallavier, France).

3.2. Plant Materials

Plectranthus scutellarioides (L.) R.Br., seeds (commercial cultivar "Arc-en-Ciel Amélioré") were obtained from Vilmorin-Mikado (La Ménitré, France). The plantlets resulting from seedlings were cultivated grew in pots (30 cm diameter and 30 cm depth), filled with a commercial garden soil (N: 250 g/m^3, P2O5: 120 g/m^3, K2O: 80 g/m^3, 37% of dry matter, 65% of organic matter, pH: 6.2, conductivity: 49 mS/cm, water retention capacity: 70% vol.) in a phytotronic room at 25°C under a 16-h photoperiod (30 µmol/m^2/s total amount of photosynthetically active radiation) and the relative humidity (RH) was approximately 30%. Plants were irrigated using overhead mist irrigation once a day and one complete watering per week. Three-month-old leaves (just before flowering in order to limit possible remobilization of RA from leaves to seeds) were collected, stored at −80°C and lyophilized prior to extraction.

3.3. Ultrasound-Assisted Extraction (USAE) Optimization

USAE was performed using an ultrasonic bath USC1200TH (Prolabo, Sion, Switzerland) with an inner dimension of 300 × 240 × 200 mm, an electrical power of 400W (i.e., acoustic power of 1W/cm^2), maximal heating power of 400W, variable frequencies, equipped with a digital timer, a frequency and a temperature controller. The sample was placed in a 100-ml quartz tube topped by a vapor condenser and was suspended in 20 ml aqueous ethanol. For this purpose, various solvent to material (S/M) ratios were evaluated (10:1, 25:1, and 50:1 ml/g DW) and various aqueous ethanol concentrations (0, 25, 50, 75, and 100 % (v/v)) were evaluated. Different ultrasound frequencies (0, 15, 30, and 45 kHz), extraction temperatures (25, 35, 45, 55, 65, and 75°C), and extraction durations (0, 15, 30, 45, 60, and 75 min) were also tested. The extract was then centrifuged for 15 min at 3,000 rpm and the supernatant was filtered (0.45 µm) before HPLC analysis.

3.4. Conventional Solid/Liquid Extraction

The conventional heat reflux extraction consisted in S/M ratio of 25:1 ml/g DW using 20 ml of pure ethanol (100%) as the extraction solvent, in a water bath with extraction temperature set at 45°C during 45, 90, and 180 min. After these extraction times, the extract was centrifuged at 3,000 rpm for 15 min, and the resulting supernatant filtered (0.45 µm) prior HPLC analysis.

3.5. High-Performance Liquid Chromatography (HPLC) Analysis

Separation was performed using Varian (Les Ulis, France) high-performance liquid chromatography system equipped with Varian Prostar 230 pump Metachem Degasit, Varian Prostar 410 autosampler and Varian Prostar 335 Photodiode Array Detector (PAD) and driven by Galaxie version 1.9.3.2 software. Hypersil PEP 300 C18 (Thermo Fisher Scientific, Illkirch, France), 250 × 4.6 mm, 5 µm particle size equipped with a guard column Alltech (Thermo Fisher Scientific, Illkirch, France), 10 × 4.1 mm was utilized at 35°C was used for the separation. The mobile phase was composed of HPLC grade solvents: A is a mixture of $HCOOH/H_2O$ at pH 2.1 and B is CH_3OH. Throughout one-hour run, mobile phase composition varied with a nonlinear gradient 8% B (0 min), 12% B (11 min), 30% B (17 min), 33% B (28 min), 100% B (30–35 min), 8% B (36 min) at a flow rate of 1 ml/min. A 10-min re-equilibration time was used among individual runs. The detection of compounds was set at 320 nm. Quantification was done based on assessment of retention time and UV spectrum of a reference standard purchased from Sigma-Aldrich (Saint-Quentin Fallavier, France). The samples examination was done in triplicates.

3.6. Experimental Design

Factorial experiment design and response surface plots were used to identify the optimal RA extraction conditions using XLSTAT2018 software (Addinsoft, Paris, France). Variables were coded at three levels −1, 0, and +1. The three independent variables and their values were selected following preliminary experiments: X_1 ethanol concentration (50, 75, and 100% v/v), X_2 ultrasound frequency (15, 30, and 45 kHz) and X_3 extraction duration (30, 45, and 60 min). Twenty-seven observations of different combinations were prepared by taking values of selected variables at different levels as shown in Table 2. Note that all independent observations were carried out in triplicate. The equation calculation as well as statistical analysis were performed as described previously [24,46].

3.7. Method Validation

The method was then validated following the recommendations of the association of analytical communities (AOAC) to ensure the precision and accuracy of the method used to quantify RA.

Five-point calibration curves were made by means of diluted solutions of RA authentic commercial standard (Sigma Aldrich, Saint-Quentin Fallavier, France). Each sample was injected three times and arithmetic means was calculated to generate linear regression equations plotting was done by the peak areas (y) against the injected quantities (x) of RA standard compound. Coefficients of determination (R^2) was used for linearity verification.

The signal-to-noise ratios (S:N) of 3:1 and 10:1, were used to respectively determine the limits of detection (LOD) and of quantification (LOQ).

Accuracy, method precision, stability, and repeatability were determined as described in Corbin et al. [26].

3.8. Antioxidant DPPH assay

The antioxidant activity was evaluated using the protocol described in [21,47].

3.9. Antibacterial Activity

The antibacterial activity was tested using the microplate protocol as described in [23,45] on *Staphylococcus aureus* ACTT6538.

3.10. Statistical Analysis

Each experiment was carried out at least in triplicate and XL-stat_2018 (Addinsoft, Paris, France) was used for all statistical analyses. All statistical tests were considered significant at $p < 0.05$.

4. Conclusions

Rosmarinic acid (RA) has already been demonstrated to be a promising compound for cosmetic applications. Nevertheless, its extraction was far from efficient. To date, many studies dealing with RA production from *P. scutellarioides* have focused on biotechnological approaches to improve its production from in planta using in vitro technologies, but little attention has been paid to the optimization of its extraction from the leaves of this ornamental plant. However, this plant is known as easy to propagate by cuttings and high renewable biomass production of leaves can be obtained through basic horticultural approaches. The development of green extraction of RA from this raw material in potential cosmetic applications is promising. Here we report on the development and validation of an efficient extraction procedure of the antioxidant and antimicrobial *trans*-rosmarinic acid from *P. scutellarioides* leaves. The present study describes an efficient and validated USAE method for RA extraction and quantification from *P. scutellarioides* leaves. A maximum RA of 11.08 % of *P. scutellarioides* leaves DW was obtained using pure ethanol a green extraction solvent in an ultrasonic bath operating at a 30 kHz frequency during 45 min. This extraction method was validated in terms of precision, repeatability, stability, and accuracy. The proposed method was proved as more efficient and less time-consuming than the conventional heat reflux method. Comparison with RA yields obtained from different *P. scutellarioides* systems confirmed that the leaves of this ornamental plant can be used as raw starting material for an efficient extraction of RA. Thus, the present method is of particular interest within the context of green chemistry in terms of reducing energy consumption and the use of green solvent. We anticipate that it could allow fast and easy extraction of RA for future cosmetic applications.

Supplementary Materials: The following are available online at http://www.mdpi.com/2223-7747/8/3/50/s1, Figure S1 Biplot representation of the linear relation between predicted vs. measured RA contents in the 27 sample extracts. Light blue contours represented p = 0.05; Figure S2 Representative chromatogram of a complete HPLC analysis of an extract of *P. scutellarioides* leaves obtained following USAE showing the presence of RA as main compound. *t*-RA: trans-RA (rosmarinic acid); IS: internal standard (4-hydroxychalcone); Figure S3 Predicted surface response plots of the antioxidant activity (% of DPPH radical scavenging activity) as a function of (a) ultrasound frequency and ethanol concentration, (b) extraction duration and ultrasound frequency, and (c) extraction duration and ethanol concentration; Figure S4 Predicted surface response plots of the antimicrobial activity (% of *Staphylococcus aureus* ACTT6538 growth inhibition) as a function of (a) ultrasound frequency and ethanol concentration, (b) extraction duration and ultrasound frequency, and (c) extraction duration and ethanol concentration; Table S1 Individual antioxidant and antimicrobial activities vs. RA contents in the 27 US extract samples.

Author Contributions: Conceptualization, C.H. and D.T.; Methodology, D.T., L.G., S.D., and S.R.; Software, S.D.; Validation, C.H., E.L., and D.T.; Formal analysis, C.H. and D.T.; Investigation, L.G., D.T., S.D., and S.R.; Resources, C.H. and D.T.; Data curation, C.H. and D.T.; Writing—original draft preparation, C.H.; Writing—review and editing, C.H., D.T., and E.L.; Visualization, C.H. and D.T.; Supervision, C.H. and D.T.; Project administration, C.H.; Funding acquisition, C.H.

Funding: This research was supported by Cosmetosciences, a global training and research program dedicated to the cosmetic industry. Located in the heart of the Cosmetic Valley, this program led by University of Orléans is funded by the Région Centre-Val de Loire.

Acknowledgments: D.T. gratefully acknowledges the support of French government via the French Embassy in Thailand in the form of Junior Research Fellowship Program 2018. LG and SD acknowledge research fellowships of Loire Valley Region.

Conflicts of Interest: The authors declare no conflict of interest.

References

1. Codd, L.E. Plectranthus (Labiatae) and allied genera in Southern Africa. *Bothalia Afr. Biodivers. Conserv.* **1975**, *11*, 371–442. [CrossRef]
2. Lukhoba, C.W.; Simmonds, M.S.J.; Paton, A.J. Plectranthus: A review of ethnobotanical uses. *J. Ethnopharmacol.* **2006**, *103*, 1–24. [CrossRef] [PubMed]
3. Namsa, N.D.; Tag, H.; Mandal, M.; Kalita, P.; Das, A.K. An ethnobotanical study of traditional anti-inflammatory plants used by the Lohit community of Arunachal Pradesh, India. *J. Ethnopharmacol.* **2009**, *125*, 234–245. [CrossRef] [PubMed]
4. Andrade-Cetto, A. Ethnobotanical study of the medicinal plants from Tlanchinol, Hidalgo, México. *J. Ethnopharmacol.* **2009**, *122*, 163–171. [CrossRef] [PubMed]
5. Kim, G.D.; Park, Y.S.; Jin, Y.H.; Park, C.S. Production and applications of rosmarinic acid and structurally related compounds. *Appl. Microbiol. Biotechnol.* **2015**, *99*, 2083–2092. [CrossRef] [PubMed]
6. Khojasteh, A.; Mirjalili, M.H.; Hidalgo, D.; Corchete, P.; Palazon, J. New trends in biotechnological production of rosmarinic acid. *Biotechnol. Lett.* **2014**, *36*, 2393–2406. [CrossRef] [PubMed]
7. Petersen, M.; Simmonds, M.S. Rosmarinic Acid. *Phytochemistry* **2013**, *62*, 43–50. [CrossRef]
8. Petersen, M.; Abdullah, Y.; Benner, J.; Eberle, D.; Gehlen, K.; Hücherig, S.; Janiak, V.; Kim, K.H.; Sander, M.; Weitzel, C.; et al. Evolution of rosmarinic acid biosynthesis. *Phytochemistry* **2009**, *70*, 1663–1679. [CrossRef] [PubMed]
9. Falé, P.L.; Borges, C.; Madeira, P.J.A.; Ascensão, L.; Araújo, M.E.M.; Florêncio, M.H.; Serralheiro, M.L.M. Rosmarinic acid, scutellarein 4'-methyl ether 7-O-glucuronide and (16S)-coleon E are the main compounds responsible for the antiacetylcholinesterase and antioxidant activity in herbal tea of Plectranthus barbatus ("falso boldo"). *Food Chem.* **2009**, *114*, 798–805. [CrossRef]
10. Falé, P.L.V.; Madeira, P.J.A.; Florêncio, M.H.; Ascensão, L.; Serralheiro, M.L.M. Function of Plectranthus barbatus herbal tea as neuronal acetylcholinesterase inhibitor. *Food Funct.* **2011**, *2*, 130–136. [CrossRef] [PubMed]
11. Falé, P.L.; Ascensão, L.; Serralheiro, M.L.M. Effect of luteolin and apigenin on rosmarinic acid bioavailability in Caco-2 cell monolayers. *Food Funct.* **2013**, *4*, 426–431. [CrossRef] [PubMed]

12. Rijo, P.; Matias, D.; Fernandes, A.S.; Simões, M.F.; Nicolai, M.; Reis, C.P. Antimicrobial plant extracts encapsulated into polymeric beads for potential application on the skin. *Polymers* **2014**, *6*, 479–490. [CrossRef]
13. Kubínová, R.; Pořízková, R.; Navrátilová, A.; Farsa, O.; Hanáková, Z.; Bačinská, A.; Čížek, A.; Valentová, M. Antimicrobial and enzyme inhibitory activities of the constituents of Plectranthus madagascariensis (Pers.) Benth. *J. Enzym. Inhib. Med. Chem.* **2014**, *29*, 749–752. [CrossRef] [PubMed]
14. Chen, Y.S.; Yu, H.M.; Shie, J.J.; Cheng, T.J.R.; Wu, C.Y.; Fang, J.M.; Wong, C.H. Chemical constituents of Plectranthus amboinicus and the synthetic analogs possessing anti-inflammatory activity. *Bioorg. Med. Chem.* **2014**, *22*, 1766–1772. [CrossRef] [PubMed]
15. Medrado, H.; dos Santos, E.; Ribeiro, E.; David, J.; David, J.; Araújo, J.F.; do Vale, A.; Bellintani, M.; Brandão, H.; Meira, P. Rosmarinic and Cinnamic Acid Derivatives of in vitro Tissue Culture of Plectranthus ornatus: Overproduction and Correlation with Antioxidant Activities. *J. Braz. Chem. Soc.* **2016**, *28*, 2017–2019. [CrossRef]
16. Qiao, S.; Li, W.; Tsubouchi, R.; Haneda, M.; Murakami, K.; Takeuchi, F.; Nisimoto, Y.; Yoshino, M. Rosmarinic acid inhibits the formation of reactive oxygen and nitrogen species in RAW264.7 macrophages. *Free Radic. Res.* **2005**, *39*, 995–1003. [CrossRef] [PubMed]
17. Osakabe, N.; Yasuda, A.; Natsume, M.; Sanbongi, C.; Kato, Y.; Osawa, T.; Yoshikawa, T. Rosmarinic acid, a major polyphenolic component of Perilla frutensis, reduces liposaccharide (LPS)-induced liver injury in D-glucosamine (D-GalN)-sensitized mice. *Free Radic. Biol. Med.* **1997**, *5*, 637–647.
18. Fallarini, S.; Miglio, G.; Paoletti, T.; Minassi, A.; Amoruso, A.; Bardelli, C.; Brunelleschi, S.; Lombardi, G. Clovamide and rosmarinic acid induce neuroprotective effects in in vitro models of neuronal death. *Br. J. Pharmacol.* **2009**, *157*, 1072–1084. [CrossRef] [PubMed]
19. Hamaguchi, T.; Ono, K.; Murase, A.; Yamada, M. Phenolic compounds prevent Alzheimer's pathology through different effects on the amyloid-β aggregation pathway. *Am. J. Pathol.* **2009**, *175*, 2557–2565. [CrossRef] [PubMed]
20. Valko, M.; Leibfritz, D.; Moncol, J.; Cronin, M.T.D.; Mazur, M.; Telser, J. Free radicals and antioxidants in normal physiological functions and human disease. *Int. J. Biochem. Cell Biol.* **2007**, *39*, 44–84. [CrossRef] [PubMed]
21. Brand-Williams, W.; Cuvelier, M.E.; Berset, C. Use of a free radical method to evaluate antioxidant activity. *LWT Food Sci. Technol.* **1995**, *28*, 25–30. [CrossRef]
22. Hano, C.; Corbin, C.; Drouet, S.; Quéro, A.; Rombaut, N.; Savoire, R.; Molinié, R.; Thomasset, B.; Mesnard, F.; Lainé, E. The lignan (+)-secoisolariciresinol extracted from flax hulls is an effective protectant of linseed oil and its emulsion against oxidative damage. *Eur. J. Lipid Sci. Technol.* **2017**, *119*, 1–9. [CrossRef]
23. Lopez, T.; Corbin, C.; Falguieres, A.; Doussot, J.; Montguillon, J.; Hagège, D.; Hano, C.; Lainé, É. Influence de la composition du milieu de culture sur la production de métabolites secondaires et les activités antioxydantes et antibactériennes des extraits produits í partir de cultures in vitro de C*lidemia hirta* L. *C. R. Chim.* **2016**, *19*, 1071–1076. [CrossRef]
24. Bourgeois, C.; Leclerc, É.A.; Corbin, C.; Doussot, J.; Serrano, V.; Vanier, J.R.; Seigneuret, J.M.; Auguin, D.; Pichon, C.; Lainé, É.; et al. L'ortie (*Urtica dioica* L.), une source de produits antioxidants et phytochimiques anti-âge pour des applications en cosmétique. *C. R. Chim.* **2016**, *19*, 1090–1100. [CrossRef]
25. Lavilla, I.; Bendicho, C. Fundamentals of Ultrasound-Assisted Extraction. *Water Extr. Bioact. Compd.* **2017**, 291–316. [CrossRef]
26. Corbin, C.; Fidel, T.; Leclerc, E.A.; Barakzoy, E.; Sagot, N.; Falguiéres, A.; Renouard, S.; Blondeau, J.P.; Ferroud, C.; Doussot, J.; et al. Development and validation of an efficient ultrasound assisted extraction of phenolic compounds from flax (*Linum usitatissimum* L.) seeds. *Ultrason. Sonochem.* **2015**, *26*, 176–185. [CrossRef] [PubMed]
27. Medina-Torres, N.; Ayora-Talavera, T.; Espinosa-Andrews, H.; Sánchez-Contreras, A.; Pacheco, N. Ultrasound Assisted Extraction for the Recovery of Phenolic Compounds from Vegetable Sources. *Agronomy* **2017**, *7*, 47. [CrossRef]
28. Bauer, N.; Leljak-Levanic, D.; Jelaska, S. Rosmarinic Acid Synthesis in Transformed Callus Culture of Coleus blumei Benth. *Z. Naturforsch. C* **2004**, *59*, 554–560. [CrossRef] [PubMed]
29. Petersen, M.; Häusler, E.; Meinhard, J.; Karwatzki, B.; Gertlowski, C. The biosynthesis of rosmarinic acid in suspension cultures of Coleus blumei. *Plant Cell Tissue Organ Cult.* **1994**, *38*, 171–179. [CrossRef]

30. Martinez, B.C.; Park, C. Characteristics of Batch Suspension Cultures of Preconditioned Coleus blumei Cells: Sucrose Effect. *Biotechnol. Prog.* **1993**, *9*, 97–100. [CrossRef]
31. Furukawa, M.; Makino, M.; Ohkoshi, E.; Uchiyama, T.; Fujimoto, Y. Terpenoids and phenethyl glucosides from Hyssopus cuspidatus (Labiatae). *Phytochemistry* **2011**, *72*, 2244–2252. [CrossRef] [PubMed]
32. Hajimehdipoor, H.; Saeidnia, S.; Gohari, A.; Hamedani, M.; Shekarchi, M. Comparative study of rosmarinic acid content in some plants of Labiatae family. *Pharmacogn. Mag.* **2012**, *8*, 37–41. [CrossRef] [PubMed]
33. Jacotet-Navarro, M.; Rombaut, N.; Fabiano-Tixier, A.S.; Danguien, M.; Bily, A.; Chemat, F. Ultrasound versus microwave as green processes for extraction of rosmarinic, carnosic and ursolic acids from rosemary. *Ultrason. Sonochem.* **2015**, *27*, 102–109. [CrossRef] [PubMed]
34. Caleja, C.; Barros, L.; Prieto, M.A.; Barreiro, M.F.; Oliveira, M.B.P.P.; Ferreira, I.C.F.R. Extraction of rosmarinic acid from Melissa officinalis L. by heat-, microwave- and ultrasound-assisted extraction techniques: A comparative study through response surface analysis. *Sep. Purif. Technol.* **2017**, *186*, 297–308. [CrossRef]
35. Adham, A.N. Comparative extraction methods, phytochemical constituents, fluorescence analysis and HPLC validation of rosmarinic acid content in Mentha piperita, Mentha longifolia and Osimum basilicum. *J. Pharmacogn. Phytochem.* **2015**, *3*, 130–139.
36. Chemat, F.; Rombaut, N.; Sicaire, A.G.; Meullemiestre, A.; Fabiano-Tixier, A.S.; Abert-Vian, M. Ultrasound assisted extraction of food and natural products. Mechanisms, techniques, combinations, protocols and applications. A review. *Ultrason. Sonochem.* **2017**, *34*, 540–560. [CrossRef] [PubMed]
37. Chatterjee, A.; Tandon, S.; Ahmad, A. Comparative Extraction and Downstream Processing Techniques for Quantitative Analysis of Rosmarinic Acid in Rosmarinus officinalis. *Asian J. Chem.* **2014**, *26*, 4313–4318. [CrossRef]
38. Tabaraki, R.; Nateghi, A. Optimization of ultrasonic-assisted extraction of natural antioxidants from rice bran using response surface methodology. *Ultrason. Sonochem.* **2011**, *18*, 1279–1286. [CrossRef] [PubMed]
39. Rakić, T.; Kasagić-Vujanović, I.; Jovanović, M.; Jančić-Stojanović, B.; Ivanović, D. Comparison of Full Factorial Design, Central Composite Design, and Box-Behnken Design in Chromatographic Method Development for the Determination of Fluconazole and Its Impurities. *Anal. Lett.* **2014**, *47*, 1334–1347. [CrossRef]
40. Amoah, S.; Sandjo, L.; Kratz, J.; Biavatti, M. Rosmarinic Acid–Pharmaceutical and Clinical Aspects. *Planta Med.* **2016**, *82*, 388–406. [CrossRef] [PubMed]
41. Ngo, Y.L.; Lau, C.H.; Chua, L.S. Review on rosmarinic acid extraction, fractionation and its anti-diabetic potential. *Food Chem. Toxicol.* **2018**, *121*, 687–700. [CrossRef] [PubMed]
42. Abdullah, Y.; Schneider, B.; Petersen, M. Occurrence of rosmarinic acid, chlorogenic acid and rutin in Marantaceae species. *Phytochem. Lett.* **2008**, *1*, 199–203. [CrossRef]
43. Snyder, L.R.; Kirkland, J.J.; Glajch, J.L. *Practical HPLC Method Development*; John Wiley & Sons, Inc.: Hoboken, NJ, USA, 1997; ISBN 9781118592014.
44. Es-Safi, N.E.; Ghidouche, S.; Ducrot, P.H. Flavonoids: Hemisynthesis, reactivity, characterization and free radical scavenging activity. *Molecules* **2007**, *12*, 2228–2258. [CrossRef] [PubMed]
45. El Abdellaoui, S.; Destandau, E.; Krolikiewicz-Renimel, I.; Cancellieri, P.; Toribio, A.; Jeronimo-Monteiro, V.; Landemarre, L.; André, P.; Elfakir, C. Centrifugal partition chromatography for antibacterial bio-guided fractionation of Clidemia hirta roots. *Sep. Purif. Technol.* **2014**, *123*, 221–228. [CrossRef]
46. Fliniaux, O.; Corbin, C.; Ramsay, A.; Renouard, S.; Beejmohun, V.; Doussot, J.; Falguières, A.; Ferroud, C.; Lamblin, F.; Lainé, E.; et al. Microwave-assisted extraction of herbacetin diglucoside from flax (*Linum usitatissimum* L.) seed cakes and its quantification using an RP-HPLC-UV system. *Molecules* **2014**, *19*, 3025–3037. [CrossRef] [PubMed]
47. Di Marco, G.; Gismondi, A.; Panzanella, L.; Canuti, L.; Impei, S.; Leonardi, D.; Canini, A. Botanical influence on phenolic profile and antioxidant level of Italian honeys. *J. Food Sci. Technol.* **2018**, *55*, 4042–4050. [CrossRef] [PubMed]

© 2019 by the authors. Licensee MDPI, Basel, Switzerland. This article is an open access article distributed under the terms and conditions of the Creative Commons Attribution (CC BY) license (http://creativecommons.org/licenses/by/4.0/).

MDPI
St. Alban-Anlage 66
4052 Basel
Switzerland
Tel. +41 61 683 77 34
Fax +41 61 302 89 18
www.mdpi.com

Plants Editorial Office
E-mail: plants@mdpi.com
www.mdpi.com/journal/plants

www.ingramcontent.com/pod-product-compliance
Lightning Source LLC
LaVergne TN
LVHW071957080526
838202LV00064B/6769